A STARTUP FIELD GUIDE IN THE AGE OF ROBOTS AND AI

Launching a startup is like climbing a mountain, just maybe more treacherous. I say this as I have spent years as a backpacker and entrepreneur. While hiking through the Alaskan Tundra, I feared brown bears and crevasses. Yet, nothing prepared me for the responsibility of payroll for over 200 families relying on my business plan to feed their children.

Unlike traditional software, the mere smell of hardware sensors and robot gearing sends shivers through most investors, with red flags arising from the perceived capital inefficiencies and intense research and development. This is coupled with a high talent requirement before launching even a minimum viable product, as these inventions demand a cross-section of skills: mechanical, electrical, and software engineering. To set out on the trail of uncrewed success, machine inventors and founders require a detailed field guide to meet customer demand and financing objectives.

My goal for this book is to help you at a pivotal point in your ideation process and, at the same time, introduce you to a cadre of potential mentors. Through interviews with some of the most respected luminaries in this field, I aim to help fortify your resolve to follow your passions and build a billion-dollar company. The chapters of this book have been organized like a field guide, as if you are setting out on a trip in the wild. Just like it's essential to satiate yourself before scaling mountains, fast-tracking your innovation into the hands of early adopters is vital for achieving success on Main Street.

Oliver Mitchell is a partner at ff Venture Capital, where he leads investments in robotics, artificial intelligence, drones, industrial automation, and climate tech while building strategic alliances with limited partners and corporate venture groups. A seasoned investor, Oliver boasts a proven track record of successful exits, including two IPOs (Novocure and Ekso Bionics), a $1.4 billion private equity sale of TripleLift, and high returns from recent transactions, including the acquisition of Scite.ai by Research Solutions (RSSS) and CardFlight's growth equity round led by WestView Capital Partners. He currently serves on the boards of CivRobotics, Cambrian Robotics, and Paraspot AI, leveraging his deep tech expertise to scale transformative companies. Previously, he built and exited ventures such as Holmes Protection, AmeriCash, and RobotGalaxy, a national STEM-focused brand. As an Adjunct Professor at Sy Syms School of Business and the author of *A Startup Field Guide in the Age of Robots and AI*, he continues to shape innovation, sharing insights on his blog as the "Robot Rabbi."

"With humility, wit, and deep expertise, Oliver Michell updates timeless entrepreneurial wisdom for the age of AI and robotics. His *Startup Field Guide* blends gripping personal stories and famous case studies with actionable advice. It feels like you're stepping into Professor Mitchell's mentoring sessions with students and founders, as he inspires and equips them to conquer the challenges of building innovative businesses."

Noam Wasserman, *Dean of the Sy Syms School of Business at Yeshiva University, and bestselling author of The Founder's Dilemmas and Life Is a Startup*

"Anyone interested in building a company in robotics should read Oliver Mitchell's *A Startup Field Guide*. The wisdom of many founders is collected and summarized in a useful and digestible form. He shows how to build a business model canvas and populate it - getting an understanding of products, customers, financing, and many of the pitfalls to avoid. The many examples of companies that have been built with different cultures and vastly different pieces of the market can help further inspire a founder to move ahead with their own ideas. An excellent read."

Howard Morgan, *Chair and General Partner of B Capital, and one of the pioneers of early-stage investing and the internet*

"With engaging writing and real-life anecdotes, this book offers readers a powerful guide to navigating the challenges and opportunities of building a business. Oliver Mitchell will give you proven strategies that will work for you 'In the Age of Robots and AI'. This book is for you, if you are looking to turn your entrepreneurial dreams into reality or scale your existing venture."

Mark Finkel, *Professor of Entrepreneurship, Sy Syms School of Business, Yeshiva University, and Founder, RightAnswers*

"This book is a comprehensive guide for aspiring entrepreneurs, providing practical insights into overcoming the challenges of building and scaling successful startups in the robotics and AI industry. Drawing from interviews with industry experts who share firsthand experiences from real-world ventures, the author distills invaluable lessons from both their successes and failures. I wholeheartedly recommend this book to my students and colleagues involved in robotics research and education as an exceptional resource for guiding their entrepreneurial pursuits."

Katsuo Kurabayashi, *Professor and Chair, Department of Mechanical and Aerospace Engineering, NYU Tandon School of Engineering*

"In this informative field guide, Oliver Mitchell shares his observations as an investor, as well as tales from across the robotics ecosystem and tips on how startups can stay on track. Mitchell surveys the numerous pitfalls facing innovators and entrepreneurs, and his workbook exercises at the end of each chapter challenge readers to apply what they've learned. Mitchell has spoken with many, if not most, of the leading robotics founders of the past two decades. Their candid recollections of their efforts, missteps, and ultimately successes are a must-read for anyone who wants to know how the industry really works. Robotics might be more collaborative than other areas of technology, but it's still very competitive as businesses try to solve some of today's hardest problems. I'd heartily recommend this guidebook to any startup, regardless of how far it has gotten in its journey from concept to commercialization!"

Eugene Demaitre, *Editorial Director, Robotics, WTWH Media*

"Oliver Mitchell is among the world's leading thinkers about robots and robotics. He is also one of the world's leading visionaries about how humans think about robots, invest in robotics, and build robotics companies. In this book, he has taken all he learned over decades of work in this field, and adapted his vast knowledge base to the Brave New World at the intersection of Robots, Humanoids, and Artificial Intelligence. The reader gets to be the beneficiary of Oliver's creation of a new knowledge ecosystem. Quick–read this before the robots do!"

Dan Burstein, *New York Times bestselling author of*
15 nonfiction books, and Managing Partner and Co-Founder,
Millennium Technology Value Partners,
a New York City-based venture capital firm

"Recent developments in AI and robotics have taken us closer than ever before to the vision of a society run by machines with little human intervention. In this book, Oliver Mitchell draws on his pioneering experiences – as a founder, management consultant, venture capitalist and teacher–to provide a step-by-step guide for navigating the challenging journey of launching and scaling a startup with these disruptive technologies. Oliver presents many case studies, practical rules, and questions for self-assessment throughout and encouraging readers to think critically about their startup ideas, adapt strategies, and build sustainable businesses. Those interested in the future of entrepreneurship, technology, or business will not regret picking up this book. While Oliver likens building a startup to a mountain expedition, reading his book is a breeze."

Deepak Hegde, *Seymour Milstein Professor of Strategy,*
Stern School of Business, New York University, USA

"In *A Startup Field Guide*, Oliver distills the wisdom, vision, and practical strategies that have made him an influential voice in the industry. This book is a great resource, offering a clear and compelling guide for founders and investors alike. Having seen firsthand the transformative potential of robotics, I can confidently say this is the definitive guidebook for navigating and thriving in this rapidly evolving space."

Fady Saad, *General Partner of Cybernetix Ventures*

A STARTUP FIELD GUIDE IN THE AGE OF ROBOTS AND AI

Oliver Mitchell

CRC Press
Taylor & Francis Group
Boca Raton London New York

CRC Press is an imprint of the
Taylor & Francis Group, an **informa** business

A CHAPMAN & HALL BOOK

Designed cover image: Oliver Mitchell with assistance from Mickey Steinerman.

First edition published 2025
by CRC Press
2385 NW Executive Center Drive, Suite 320, Boca Raton FL 33431

and by CRC Press
4 Park Square, Milton Park, Abingdon, Oxon, OX14 4RN

CRC Press is an imprint of Taylor & Francis Group, LLC

© 2025 Oliver Mitchell

ISBN: 978-1-032-83247-0 (hbk)
ISBN: 978-1-032-82749-0 (pbk)
ISBN: 978-1-003-50841-0 (ebk)

DOI: 10.1201/9781003508410

Typeset in Sabon
by SPi Technologies India Pvt Ltd (Straive)

For Rabia,

*And in appreciation of all the innovators
mentioned herein.*

CONTENTS

LIST OF INTERVIEWS (IN ORDER OF APPEARANCE)

The first rule of leading a startup is to hire people smarter than yourself. This adage coined by Steve Jobs has guided me in everything from founding companies to funding disruptive technologies. As a practice, I am less than six degrees away from anyone I want to meet. When embarking on a Field Guide, I turned to the experts I have admired and worked with over the years – these are my teachers. The interviews described below and chronicled on the following pages are firsthand accounts of the lessons learned from some of the most well-known and forward-thinking luminaries in the world of mechatronics. These founders, inventors, scientists, investors, and policymakers are your sherpas, leading you to the peak of your mountain.

- Mick Mountz, founder of Kiva Systems (acquired by Amazon in 2012)
- Dr. Yannis Ieropoulos, creator of Microbial Fuel Cells
- Helen Greiner, co-founder of iRobot and CyPhy Works
- Joe Jones, inventor of iRobot's Roomba and co-founder of Tertill
- Paolo Pirjanian, former CTO of iRobot and founder of Embodied
- Samantha Lee, founder of Meili Technologies
- Erik Nieves, co-founder of Plus One Robotics
- Lior Susan, founder and managing partner of Eclipse Ventures
- Dor Skuler, founder and CEO of ElliQ
- Tom Stevens, inventor of Tombot
- Abe Murray, general partner of AlleyCorp, formerly of Google
- Tom Yeshurun, co-founder and CEO of CivRobotics
- Jerry Sgobbo, founder of Dive Technologies
- Ryan Gariepy, co-founder and CTO of Clearpath Robotics (acquired by Rockwell Automation)

- Rodney Brooks, Panasonic Professor of Robotics (emeritus) at MIT, and co-founder of iRobot, Rethink Robotics, and Robust.AI
- Andrea Thomaz, founder and CEO of Diligent Robotics
- Atle Timenes, founder of wheel.me
- Charlie Andersen, founder and CEO of Augean Robotics
- Daniel Theobald, founder and CEO of Vecna Robotics
- Sean Simpson, CVC investor formerly with Yamaha Motors and GM Ventures
- Kara Jones, executive director of GENIUS NY
- Professor F. Javier Diez of Rutgers University and founder of SubUAS
- Retired Colonel Patrick Mahaney, US Army
- Dr. Henrik Christensen, author of A Roadmap for US Robotics
- Tom Ryden, executive director of Mass Robotics
- John Ha, founder and CEO of Bear Robotics
- Raffaello D'Andrea, co-founder of Kiva Systems and CEO of Verity
- Nic Radford, formerly of NASA, founder of Nauticus and Persona AI
- Hailey Nichols, founder and CEO of Locus Lock
- Claus Risager, founder of Blue Ocean Robotics

INTRODUCTION

Adventure and Risks of Startups: Trailhead

Preparing for the Summit

Launching a startup is like climbing a mountain, just maybe more treacherous. I say this as I have spent years as a backpacker and entrepreneur. While hiking through the Alaskan Tundra, I feared brown bears and crevasses. Yet, nothing prepared me for the responsibility of payroll for over 200 families relying on my business plan to feed their children. At the same time, before I embarked on a trek in the backcountry, I took great pains to plan my provisions, clothing, routes, and emergency protocols.

However, too often, my startups moved at supersonic speeds that forced carefully laid out plans to be thrown out and replaced with pivots without much forethought (essentially, there is a moment when all founders are flying by the seat of their pants, even though there is a hole in their tush). The result is a PowerPoint strategy, bullet points instead of detailed customer discovery and feedback, leading to rash decisions and misspent dollars. Now if this is true for all startups, robot companies are amplified by the sheer costs of capital, sensors, and talent. Just like setting out on an adventure in the woods, a field guide is required to determine the best course of action.

Learning from Adrian, the Romanian

I heard this story at 12,000 feet in the middle of a whiteout while staring up at Denali, the highest mountain in North America (20,320 feet), and it has guided me through every venture (both outdoors and indoors). One of our climbing guides shared an article from Backpacker Magazine 1991 about Adrian, the Romanian climbing legend at Mt. Rainier. The Eastern European

DOI: 10.1201/9781003508410-1

backpacker set out one summer to climb the Alaskan peak. Denali is one of the highest ascents out of all the major world peaks, 18,000 feet from the base, and, as such, it also has one of the highest rates of evacuations in the US National Park system. To avoid the expense and loss of life, Park Rangers instituted a strict policy of meeting all climbers to review the trek and provisions with backpackers before setting off.

Adrian arrived a bit late in the day to meet the Rangers at the Talkeetna Station, but still, they asked to see the contents of his backpack. Surprised, the Rangers stopped Adrian and asked how he was going to survive with one stove. Glacier living requires at least two stoves, one for melting snow for water and another for cooking food. The bold Romanian boasted that he was "going to buy water from the other climbers." Sure enough, he was evacuated days later, almost freezing to death. The following year he summited the mountain with two stoves. His perseverance and survival instincts in listening to valued advice and implementing it are the attributes that we look for in budding founders.

Robot Startups Are Different

Thomas Edison once commented, "Genius is one percent inspiration and 99 percent perspiration." In the world of venture capital, such sage advice rings even louder. This is especially true for robotics and AI startups. Unlike traditional software, the mere smell of hardware sensors and mechanical gearing sends shivers through most investors with red flags arising from the perceived capital inefficiencies and intense research and development. This is coupled with the high talent requirement before launching even a minimum viable product as these inventions demand a cross-section of skills: mechanical, electrical, and software engineering.

Finally, the sales cycles tend to be slower for mechatronics due to the nature of disrupting conventional methodologies, such as collaborative robots displacing human hands, autonomous vehicles taking the wheels from drivers, and drone missions soaring over pilots and site inspectors. At the same time, when it works, it pays huge dividends with the success of Apple, Tesla, and DJI, to name a few billion-dollar companies. To set out on the trail of mechanical success, machine inventors and founders require a detailed field guide to meet customer demand and financing objectives.

After founding RobotGalaxy in 2006, I became known in the New York ecosystem as "Oliver the Robot Guy." Then in 2011, with the popularity of blogging, my marketing department created a site for me called "The Robot Rabbi." The reasoning was I'm Jewish, so naturally I would enjoy the promotion. Putting aside the religious title, a rabbi at the end of the day is a teacher. So as I stared at the blank page for my first post, I decided to become a teacher of robots for my fellow founders and investors. Who knew that my voice filled a void in the industry? Most articles then focused on technology and engineering, so I questioned the business applications and models.

This led me to meet hundreds of amazing innovators and see thousands of business plans in the industry. Eventually, I organized an investment vehicle called Autonomy Ventures to seed mechatronic startups. Fortunately, my efforts paid off with eight exits over six years, including two public offerings and one unicorn private equity sale. The "cleric" in me started documenting the secrets of these successful ventures, as I continued to chronicle the industry on my blog and trade publications.

Employing my pastoral made-up title, indulge me in sharing a biblical thought. The sages say that Moses was the humblest man ever to live. Still, at the same time, he was a motivational leader, successful innovator, and product visionary in establishing the Torah. In interviewing thousands of inventors for my blog, in addition to the more than 20 luminaries in the field of robotics highlighted in this book, I think Daniel Theobald, founder of Vecna Robotics, has a way of reconciling the two sides of Moses that all founders could harness when launching their ideas.

Theobald shared a recent encounter with a group of entrepreneurs at one of his workshops: "One of the founders asked me, 'When you were in our position would you take all this advice?' I said absolutely not because I thought I knew everything. But you know, I guess in retrospect, I would say, ask for help." He continued to unpack his statement, "Being a founder is one of the most challenging things. But ask for help. That's what a smart founder would do. Build a network of knowledgeable people. I found over the years two or three mentors that I think made me way smarter."

My goal for this book is to help you at a pivotal point in your ideation process and, at the same, introduce you to a cadre of potential mentors (albeit in print) who can fortify you in following your passions and building a billion-dollar company. The chapters of this book have been organized like any field guide, as if you are setting out on a trip in the wild. The key to packing for a trip is understanding the conditions and environment you will be facing before setting out. In business terms, this is the startup ecosystem, which varies according to the type of solution you aim to solve. Along the way, the field guide will also introduce other facets of the ecosystems that could advance your invention in anticipation of raising private capital.

I have also provided interactive elements for you to quiz yourself on mastering the topics discussed by evaluating some of the most audacious ideas I've published over the years on my blog. Each workshop will include an assignment activity to help you pump your business muscle before embarking on the next chapter with a short exercise introducing the coming material (also introducing a new facet of the innovation ecosystem). At the end of this handbook, I've also included a lexicon of critical terms that every founder (and investor) should familiarize themselves with before venturing on a startup climb. Just like it's essential to satiate yourself before scaling mountains, fast-tracking your innovation into the hands of early adopters is vital to Main Street implementations. The first step starts with reading the map, understanding the topography, and charting your course. Let's begin!

1

CUSTOMER DISCOVERY + PRODUCT MARKET FIT

Planning

In 2012, Amazon.com acquired Kiva Systems for $775 million, making it one of the largest purchases of a robot startup ever.[1] However, the headlines failed to mention that the idea for Kiva came out of one of the most high-profile dot-com flops. In November 1999, Webvan had all the ingredients of a high-flying dot com startup, an initial public offering (IPO) – an all-star management team, a huge market, a disruptive business model, and a seemingly endless supply of cash (including $375 million of new capital).[2] The only thing missing was real customer validation with only $395,000 in revenues on top of $50 million in losses. Top-tier investors, like Benchmark Capital and Sequoia Capital, remained confident in their early bet. Fast forward less than two years later, and the online grocer declared bankruptcy.

The startup followed the advice of its board to "get big fast" and drive home its "first-mover advantage" early. Their $800 million mistake was following the tried-and-true path of a big company that follows a linear launch plan from concept to product development to beta testing to shipping. It sounds sane, but such models in today's fast-paced world of online businesses do not leave room for pivots or product interactions. As Steve Blank recommends in his book *The Startup Owner's Manual*, the product development process is a customer development process that is circular, looping back to the ideation stage with early feedback for redesigns, redos, and pivots (see graph below from Blank's book)(Figure 1.1).

DOI: 10.1201/9781003508410-2

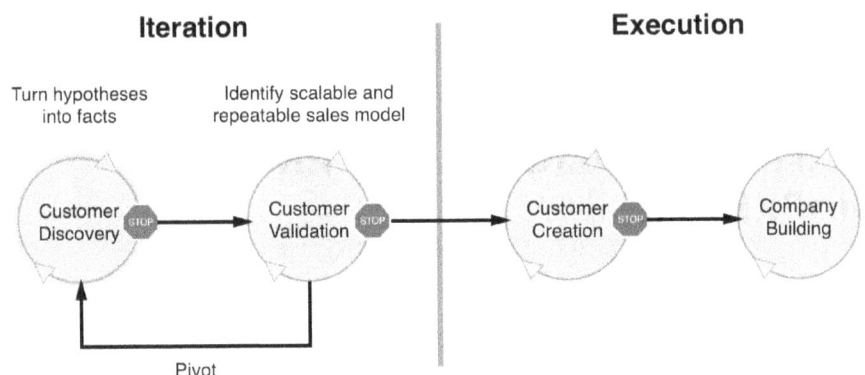

FIGURE 1.1 The customer development process.

Credit: The Startup Owners Manual, 2012.

Kiva Systems: Early Market Validation

In 2003, a former Webvan executive, Mick Mountz, thought there might be a better approach to automating the e-commerce pick-pack-and-ship process using mobile robots. He founded Kiva System to pursue that idea.

"At Webvan, my role was director for the business process of logistics. When I arrived in the fall of 1999, it was functioning and delivering orders, but it was losing money on every order, and it was very labor intensive. So in my first three months on the job, I did a lot of looking into the cost structure and how the orders were being filled. I quickly came to the realization that 70% of the cost of running those warehouses was manual labor and 70% of the direct labor's time was in walking around the warehouse to find the items for these orders. I'm thinking 49% of all the costs of the order is for paying people to walk around," Mountz explained about Webvan's inefficiencies.

Recognizing the business pain point as Mountz did at Webvan to birth Kiva is a crucial first step for any founder. Mountz experienced firsthand the broken economics of Webvan. His co-founder, Dr. Raffaello D'Andrea, reflected, "Mick had a great quote. He said that they were taping the equivalent of a $20 bill on every order that went out the door 'cause their costs were so high."

Mountz's customer discovery process started out while employed by the online grocery dot com. Empowered with a mission to reduce operating costs, Mountz began to evaluate all the other offerings in the industry for Webvan. He remembered, "I started interviewing material handling companies in the industry because my job was to figure out the next [warehouse] building. Each company that I interviewed you know they kind of had a hammer looking for a nail." He continued to paint a picture of how frustrating it was for early e-commerce companies.

"So if you talk to a tilt-tray vendor, they would say tilt-tray systems are state of the art, everyone is using them. I would ask how it would work with a jar of Prego, for example. And they said, well, you can't slide glass jars down the chute. He went through countless examples of liters of soda, round rollable items, eggs, and fresh fruit, and in the end, there was no off-the-shelf, ready-made solution for online fulfillment, which required pick-and-pack of individual items versus full handling of full cartons," he reminisced.

It was around this time that two things happened for Mountz. One, he left Webvan as it was closing fulfillment centers and heading toward bankruptcy, and two, he had an epiphany. In his words, "I was still thinking about the problem, with all that walking and now with the emergence of mobile robotics, I started putting two and two together. What if we could move the products around, instead of moving the people?" Kiva was born.

Mountz then shared with me his process of setting up Kiva. He said, "We just started kind of a race to get the intellectual property. So, the very first thing I did in the spring of 2002 was file a couple of patents on the idea of mobile robots picking up mobile inventory pods. Nobody had described that before." Once the idea was legally secure, the young entrepreneur triggered a customer discovery journey by talking to people at the Gap, Motorola, J. Crew, Dell, and a handful of others.

"I would go to them and say, what if you had a system with mobile robots that could move stuff around and bring the products to your pick workers? They would say with an incredulous tone sure, if you had something like this, maybe it would work for us. Come back when you have something," described Mountz.

After some initial market validation, Kiva's founder eventually filed his incorporation papers and raised a friends and family $150,000 seed round. This enabled him to build a working prototype and capture a video illustrating how a human picker can pack goods at his/her station with inventory arriving on a moving pod.

Now empowered with a proof of concept, Mountz reminisced, "I went back to those same customers, and said this is what I was talking about last year, and they said, Oh, I see what you're talking about now."

Eventually, after closing some additional capital, Kiva was able to build a complete prototype system to convey the concept at a larger scale. Mountz shared, "We built a small system with about six spots and 16 pods, and we had customers like Staples and Walgreens, and others would come by our offices and check it out. We then leveraged those meetings into pilot conversations that became permanent purchase orders for full-scale systems."

Tip$: Turning Pilots into Purchase Orders

In my job as a venture capitalist, I see numerous pitch decks each with big logos listing the pilots of the startup. At the same, I'm amazed by why so few pilots never turn into recurring revenue streams, especially for robots. Sometimes the reason is bureaucratic with innovation departments removed from business units, or other times it's a warning sign to investors that the great innovation is mismatched and performs poorly.

Speaking to Mountz, I pressed him on Kiva's superpower for generating meaningful sales from early feasibility tests. "So we always wanted to be clear early on, 'What's the purpose of the pilot? What are you trying to learn from the pilot?' And then we would put those metrics in the contract," Kiva's founder asserted.

He illustrated this from his first customer, Staples. They retorted to Mountz's questions, "We want to see if this thing can actually do 300 units an hour."

Mountz responded, "Okay, so we will do it, is there anything else?"

Stapes pushed him further, "We want to see if the system can run all day, you know, a ten-hour shift without needing a lot of interventions from technicians or anything."

Now with metrics defined, he asked, "If we can prove all of those things, what's the next step? If we're successful on this pilot in improving these metrics, does that give you everything you need for a business case to buy a million-dollar or a five million dollar Kiva system?"

The customer, with millions on the line, would then add questions on spare parts and servicing. This forced the return-on-investment (ROI) calculations to be tabulated early on by both parties, at the same, Mountz affirmed that his approach enabled him to go from pilot to sale with "pre-wired" key performance indicators (KPIs) that their internal champion could plug into make the business case and go from tester to purchaser fairly seamlessly.

To underline this point, Mountz boasted that over the seven years from his initial financing to 2011, Kiva had 12% of the top 100 e-commerce companies as clients, including Diapers.com[3] and Zappos,[4] which were later acquired by Amazon (and eventually led to the Kiva acquisition) (Figure 1.2).

"It's About the Business Problem, Not Tech"

Today, Mick Mountz offers advice to entrepreneurs looking to follow their passions. In prodding him for his first principles that every robot company must follow, he told me the following anecdote: "When I give talks at startup boot camps, I'm always reminding folks it's not about the technology. It's about the business problem you're solving. So many teams want to just build a cool widget, and I always say, 'Work backward from the customer problem you're trying to solve.'"

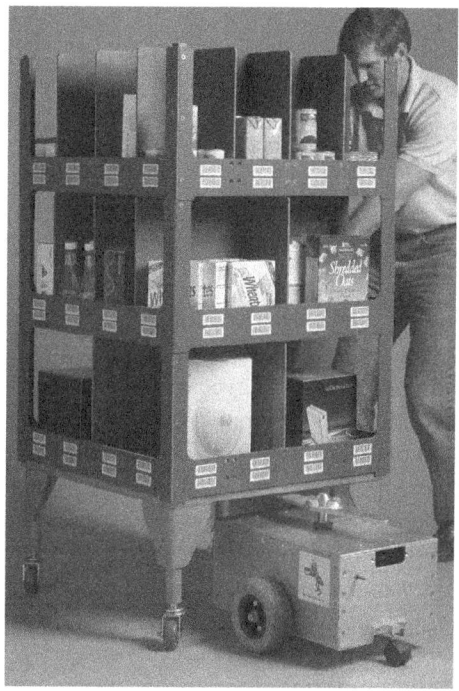

FIGURE 1.2   Mick Mountz testing an early Kiva prototype.

Photo credit: Kiva Systems, 2003.

Mountz's own experiences back up this belief as he set out to build an autonomous warehousing platform, but after experiencing the inefficiencies of Webvan firsthand and speaking to countless users, he fell upon robots.

Mountz instructed, "Don't get overly enamored with the technology. Build the technology that meets the solution. Stay lean, stay focused. Stay close to the customer." He further emphasized, "In the case of Kiva, we wanted a business where our fate was in our own hands, so we did the entire solution end-to-end. We designed it, built it, sold it, installed it, maintained it, and supported it." He added, "Something I always coach startups on is just be careful with outsourcing, understand what and why, and the pros and cons of doing that." He stressed that what made Kiva successful was its reliability: "A lot of times there was pressure to use somebody else to build the robots, but we resisted that because we knew that our success depended on the customer experience being seamless from end-to-end."

Technologist Paolo Pirjanian, who I will talk about more in the next chapter, coincidentally echoed Mountz's advice, as he outlined the first questions one has to ask about a robot/AI venture: "Practically speaking, you want to understand what is the problem you're trying to solve? What is the pain point you're trying to address, and is that pain point really a pain point? Is it a big enough pain point? And is robotics the right solution for it, because many times this is where robotics fall apart, and why robotics have not been adopted as rapidly as we have thought it would be."

Pirjanian further cautioned, "The game is changing from now moving forward. But I think the last 20 years is because many times the solutions that we think of as technologists are more of a technology push. And those companies tend to fail. Big time. Whereas you want to better start with understanding the market that you're going after. What is a pain point you're trying to address, and is the solution you have in mind the right approach to solving it? Or is it mostly because you're enamored by the technology."

Pirjanian advised that once a startup finds product market fit after a thorough customer discovery process, it's time to "Fit as fast as you can because where robotics gets complex is when it meets the real world." A robot both utilizes data and acquires data through usage. He further warned, "The challenge is that the real world is extremely complex. There is a lot of noise and ambiguity like if you're using a camera-based solution what do you do if you're in a dark room where you can't see anything?"

At the same time, it's important to note that the customer doesn't care how complex the problems are, as he clarified, "The customer cares about the fact that they paid money to buy a certain benefit, and you better deliver that no matter what the environmental conditions are." The roboticist stresses the importance of fielding or testing robots outside of simulations or lab environments early is critical. To quote, "It's very easy to make things work in a lab, but as soon as you step outside of the lab that's where things get exponentially difficult."

Questions To Ask Yourself Before Launching*

1 What is the business problem you are solving, and why?
2 What is already in the marketplace, and how is it broken?
3 Why is the proposed technology the best way of solving the problem?
4 What are the economics for the end user to adopt your solution?
5 Who are the buyers, and how much will they pay for it?

*Note: You can only ask yourself these questions after conducting a thorough customer discovery process that includes speaking with at least 500 prospective users. At any time, if any of the above answers are negative, iterate and evolve your idea until 10% of your potential buyers scream, "If you build it, I will take it!"

Founder Insights: Yannis Ieropoulos, Microbial Fuel Cells

Mick Mountz of Kiva experienced the pain point firsthand of online fulfillment from the early days of e-commerce. Professor Yannis Ieropoulos of the University of Southampton in the United Kingdom has developed a novel process to tackle another problem: climate change.

Background

The World Bank estimates that humans produce 3.5 million tons of solid waste daily, with America accounting for more than 250 million tons a year or over 4 pounds of trash per citizen.[5] This figure does not include the annual 34 billion gallons of human organic materials processed in water treatment centers nationwide.[6] To the fictional Dr. Emmett Brown, this garbage is akin to "black gold" – ecologically powering cities, cars, and machines. In reality, the movie "Back to The Future II" was inspired by the biomass gasification movement of the 20th century in powering cars with wood during World War II when petroleum was scarce. The technology has advanced so much that a few years ago, the GENeco water treatment plant in the United Kingdom built a biomethane gas bus that relied solely on sewage. In reflecting on the importance of the technology, Collin Field of Bath Bus Company declared, "We will never, ever, ever, while we are on this planet, run out of human waste."

Invention

In May of 2021, Dr. Ieropoulos demonstrated his revolutionary Microbial Fuel Cells (MFCs) to me. As witnessed, he is not just inspired by nature but harnessing its beauty to power the next generation of robots. The MFCs mimic an animal's stomach, with microbes breaking down food to create adenosine triphosphate (ATP). His lab began building MFCs to power its suite of EcoBots. "I started this journey about twenty years ago with the main purpose of building sustainable autonomous robots for remote area access," reflected Ieropoulos.

Ieropoulos' team started with the idea of using MFCs to power machines by putting the microbes directly inside the unit to more efficiently produce energy from any sugar-based substance, even waste (e.g., urine, feces, and trash). The Professor described the elegance of his technology that creates a "uniform colonization" of microbes multiplying every 8 minutes with parent lifecycles succeeded by daughter cells in a continuous pattern of 'feed-growth-energy.'

He compared it to the human microflora process of breaking down fresh food in the digestive system that results in healthy bathroom visits, "The same is with Microbial Fuel Cells; as long as we continue feeding them, the MFCs will continue to generate electricity." The professor boasts that batteries perform better than anything on the market as biological lifeforms have no denigration since "The progeny keep refreshing the community or electrodes so we have stable levels of power." By contrast, the most popular non-fossil fuel available, lithium, degrades over time and leads to destructive mining practices in scarring the Earth in search of declining ore resources with the explosion of mobile phones, portable devices, and electric vehicles.

Business Plan

While MFCs are still in their infancy, Ieropoulos shared his plan for commercializing the invention. His lab recently announced the success of its MFCs prototypes in powering mobile phones, smartwatches, and other devices (including the EcoBots). In addition, Ieropoulos has pushed his team to miniaturize the size of his batteries from its 6″ prototype to smaller than an AA while at the same time rivaling the performance of alkaline. Backed by the Gates Foundation, he reduced production costs from $18 to $1 a unit before achieving economies of scale with mass production.

His strategy has expanded beyond autonomous robots to power smart homes by connecting multiple MFCs to a house's sanitation and waste systems. "Our research is all about optimizing miniaturization and stacking them with minimum losses so we can end up with a car battery-like shape and size that gives us the amount of power we require," explained the professor. When I questioned Ieropoulos about using MFCs in the future of autonomous fleets and even to offset the high energy demands of something like bitcoin mining, he remarked, "It would be naive of me to say a straight yes, but this is, of course, the work we are doing. I strongly believe the development of new materials will help with the energy density." In thinking more about his work, he declared, "We have yet to see the full potential of Microbial Fuel Cells. I do think one day we will have a 'Back to the Future' scenario, feeding your food scraps to your car."

Pressing Ieropoulos on how he envisions robot-charging stations working in a factory or home soon, he illustrated it best, "With a Roomba example, it actually picks up food scraps in the kitchen that would be a very nutritious source of fuel for the Microbial Fuel Cell, but that's a few steps down the line." He continued, "A straightforward application for something like a Roomba is to leave the charging station where it is and connect it to the toilet or kitchen sink. The fuel cycle would be continuous as the robot would not be drawing energy from the house, but the wastewater."

✍ Assignment

While innovations like MFCs are part of a new wave of inspirational, environmentally focused solutions, customer adoption depends ultimately on executing a plan. Applying the questions outlined in this chapter, what steps would you recommend the Professor take to run a robust customer discovery process and validate the commercial viability of his ingenious solution? Use the MFC exercise to flex your biz pitch muscles.

Notes

1 Guizzo, Erico. 2012. "Amazon Acquires Kiva Systems for $775 Million - IEEE Spectrum." Spectrum.ieee.org. March 19, 2012. https://spectrum.ieee.org/amazon-acquires-kiva-systems-for-775-million.
2 Emert, Carol, and Chronicle Staff Writer. 2001. "Venture Lessons in Webvan Collapse / Financing History a Cautionary Tale." SFGate. July 15, 2001. https://www.sfgate.com/bayarea/article/Venture-lessons-in-Webvan-collapse-Financing-2899418.php.
3 Yarow, Jay. 2010. "Amazon Announces $545 Million Acquisition of Diapers.com." Business Insider. November 8, 2010. https://www.businessinsider.com/amazon-announces-500-million-acquisition-of-diaperscom-2010-11.
4 Lacy, Sarah. 2009. "Amazon Buys Zappos; the Price Is $928m., Not $847m." TechCrunch. July 22, 2009. https://techcrunch.com/2009/07/22/amazon-buys-zappos/.
5 McDonald, Juliana. 2023. "Curbing America's Trash Production: Statistics and Solutions." Dumpsters.com. February 17, 2023. https://www.dumpsters.com/blog/us-trash-production.
6 US EPA,OW. 2019. "The Sources and Solutions: Wastewater | US EPA." US EPA. April 18, 2019. https://www.epa.gov/nutrientpollution/sources-and-solutions-wastewater.

2

TEAM CULTURE DRIVES INNOVATION

Base Camp

Great companies don't just create innovative products. Great companies build an amazing corporate culture that reflects the unique attributes of their brand; robotics companies are no different. Reed Hastings, founder of Netflix, calls his streaming service a "culture of reinvention." This catch-phrase rings very true for Netflix watchers. The startup launched from a mail-order business to an entertainment powerhouse with an award-winning producer of original content and international market domination. The speed with which the company has been able to move and redefine the movie industry is groundbreaking. Hastings attributes this success not to his genius but to his ability to attract great talent (essential for any mechatronic startup). I highly recommend his book, *No Rules: Netflix and the Culture of Reinvention.* In looking at hardware, iRobot's path and early iterations are an excellent source of inspiration.

iRobot: Hard Lessons Learned Early On

The story of iRobot, one of the early robot success stories, is one of reinvention before dominating the autonomous vacuum cleaner market, spiking at 75% market share.[1] Spun out of MIT Artificial Intelligence Lab, two post-graduate students, Colin Angle and Helen Greiner formed the startup with Professor Rodney Brooks in 1990. As Greiner described, "We didn't start with a business plan. We started with a desire to build robots."

In many ways, the early engineers took the opposite path discussed in the previous chapter of embarking on a diligent customer discovery process before launching. This could be why the trio started by offering professional

DOI: 10.1201/9781003508410-3

services to the government and local manufacturers, including Rhode Island headquartered toy behemoth Hasbro. Unfortunately, while mechanical babies paid the rent (sometimes), it would be almost a decade before the team would pivot from project-based revenue to standalone products.

As the former president and chair of iRobot, Greiner stated, "It was 12 years before we put the Roomba on the market. But you know, we were basically just out of grad school. And it wasn't like today when there are entrepreneurship classes where you can learn this stuff. We really graduated out in the world trying to start a company. We had to run everything on our own. We went kind of hand-to-mouth for many, many years as if we didn't have the resources to pay payroll at the beginning of the month."

As she explained, it was a cautionary tale for any startup founder looking to forgo a thorough customer discovery process, "It's not a very good startup story for your book, because it's not the right way to start a company. But we have learned that since." Greiner continued to elaborate how in the early days she would cover expenses and stay afloat, "Cash flow was very, extremely tough. We weren't capitalized. And basically, building any robot for money. We had deals with large companies that you would know. We would bring technology and creativity. Now we had some great prototypes, even products on the market. But they didn't go anywhere, and I think it's because the large companies just go at a glacial pace."

The frustration of running a services business boiled over as iRobot's audacious dreams were held ransom by big corporate decision-making. Greiner now expressed, "The one thing that people [founders] need to realize is if you're talking to one person there, they're not necessarily speaking for the entire company. And then, if that one person gets promoted, it ends up sideways. They moved companies, switched jobs, whatever, which can affect the whole deal, you're kind of orphaned in the company."

Her cynicism of multinationals led iRobot to make a great leap of faith, as she declared, "Even if you've got a million-dollar deal signed, or even multiple million-dollar deals, it doesn't matter to large companies. We assumed that they invested a few million dollars in something, and we were going to take it to the finish line. But that wasn't the case at all; they were just wetting their feet and keeping tabs on technology rather than getting a product on the market. So, we found as a small company, it was better for us to get a product on the market under our own brand and be able to take it on ourselves."

iRobot's Product Evolution

As iRobot transitioned from a consultancy shop led by engineers to a venture-back product company, two ideas gained traction inside the company. The first one was a military robot called PackBot. This grew out of the government services business post-9/11. PackBot was a remote-controlled robot for detonating improvised explosive devices (IEDs), crucial for the then-growing conflicts in Afghanistan and Iraq.[2] The second was a riskier bet, a consumer robot vacuum cleaner, Roomba. The invention was born out of cocktail discussions that one of iRobot's first employees had at the time.

As Joe Jones, its creator, elucidated, "It went like this, every time I would go to a party where I was introduced to new people, I would say, I'm a roboticist. And they would say, Oh, that's so cool! 'Can you build me a robot that will clean my floor?' It was uncanny how that almost always happened, and it didn't happen just to me, it happened to other roboticists as well." As it turns out, vacuuming the floor is high up on the consumer's wish list in the world of dull, dirty, and backbreaking work.

It would take a few years before Jones' dream of Rosie from *The Jetsons* would come to life. He recalled, "So that's where we left it for like 8 years, after lots of lots of fun, robot projects, no big financial breakthroughs. Gradually, our robot company got to the point where they could contemplate doing a development like Roomba." It took a professional services contract for a large cleaning robot to prompt a Jones colleague to suggest, "'Why don't we build a little robot that will clean floors?' and he thought we should try again." The product team decided that the first step was a prototype to convince management of the idea. "We showed the prototype to Colin and the others, and they decided on the spot: 'This seems like a good idea,'" boasted Jones (Figure 2.1).

FIGURE 2.1 Robot vacuum prototype (in Lego) for the 1989 MIT AI Lab Olympics.

Photo credit: Joe Jones, 1989.

It was December 1999, when Internet excitement filled the air with the recent high-flying initial public offerings of Yahoo!, Google, and eBay, among others.[3] This represented a challenge for the hardware company seeking the necessary capital to launch its new Roomba-led strategy.

As Greiner described, "We started raising venture capital. Now it wasn't easy to raise venture capital then because I would get to a meeting and then would say, 'Oh, you don't do Internet. We don't have time for this.' But we started raising capital just $1 million I think in '98, and then $1.5 [million] in '99, and $5 [million] and $7 [million] in the next few years. That allowed us to invest in our ideas. And I think that's where it comes to the culture that we had created."

In reflecting, Greiner shared, "I think the culture did greatly enable the Roomba to develop and the hard work of the team. We kicked it over the goalposts! It takes concerted effort where you put everything aside until it gets done...I'll probably remember that as the best time of my life."

Acquiring Talent: Roomba Becomes Intelligent

As the iRobot team defined a new category for automated home appliances, an Armenian roboticist, named Paolo Pirjanian, left his dream job at NASA to join the infamous dot com entrepreneur Bill Gross at Idealab. This journey eventually led Pirjanian to Boston to become the chief technology officer (CTO) of iRobot in 2008 (after Greiner stepped down). Before doing so, he partnered with Bill Gross to percolate a robotics startup that pushed the envelope of what was possible during the turn of the twenty-first century.

As he recollected, "I was there doing my childhood dream job, and then I got contacted by Bill Gross of Idealab. And initially, I just rejected it. Then I started looking at Bill, and I saw that he had built multiple very successful companies during the dot com era. And then I got really curious about entrepreneurship, how you build companies, how you raise money, and all that." In speaking with Gross, Pirjanian was blown away by his grandiose plans of building the "Microsoft of Robotics." While skeptical of the particularity and sophistication of robots in 2001, he set off with Gross to create Evolution Robotics.

In speaking with Pirjanian, his experiences at Idealab shaped a playbook that he would take in all his future endeavors in the industry (as paraphrased below):

1 **Complexity**: engineering robotics is extremely difficult, so important to recognize the obstacles and avoid them;
2 **Cost of Goods**: one key constraint when building a commercial product is cost. Aligning your unit economics with your value proposition is critical

(unlike in academia, where it's about building a demo to publish a paper, and cost is not the primary issue).

3 **Reliability**: imagine shipping thousands, or millions, of units of a product that needs to work in a location you have never seen. In the case of Roomba, someone comes home from Best Buy and presses a button, and it has to work out of the box regardless of the environment.

4 **Capital**: understanding the challenges of convincing investors (especially venture capital) of a complex startup that encompasses hardware, software, distribution, manufacturing, etc.

Pirjanian illustrated his last point by comparing his experiences back in the early 2000s to today: "Believe me, when you would walk into a room on Sandhill Road [where all the VCs in the Valley are located] to pitch, as soon as you said hardware and robot, that was the end of the conversation. Now it's very different."

It was this philosophy, with the support of Bill Gross, that led Pirjanian to build a cost-effective computer vision simultaneous localization and mapping (SLAM) technology that was eventually acquired by iRobot to enable the Roomba to map everyone's home. Following his second principle regarding unit cost, Evolution Robotics introduced an autonomous navigation solution for hundreds of dollars that would typically cost tens of thousands of dollars back then. Eventually, he would bring down the costs to less than $10 a unit before being acquired by iRobot.

"If a robot vacuum cleaner costs $2,000, no one is going to buy it because people are comparing it to a $50 to $100 upright vacuum cleaner. So, you better be in the low hundreds. Understanding that at the end of the day you have to make money, that your customers have to make money, and the price has to make sense relative to the value-to-price ratio for the end users is crucial. So, the biggest lesson for me repeatedly in the last 20 years is price, price, price, price!" stressed the former CTO.

Similar to Greiner, Pirjanian attributed his success in achieving cost-effective production to the quality of the iRobot team: "The enabler was to find the right people who know how to get stuff done. A team that really understands how to maximize the technology." He illustrated with an example, "We had developed this algorithm that was running on a say a $50 processor in terms of compute power and then camera and other logic board. It would come to about a hundred dollars. What we needed to do was get it to run on a $5 processor. So, you needed to have literally the best experts in the world to understand how to optimize the code and find the right trade-offs to do that as well."

He further professed, "They say necessity is the mother of invention, which is absolutely true. For months I couldn't sleep, as I was thinking about ideas of how we would figure out how to solve this problem. And I came up with not only one, but two ideas, and both worked." He sums up the impor-tance of a good team as a "key enabler, plus a lot of creative thinking when you're faced with a challenge."

Being a Talent Hive

Pirjanian whispered his secret to luring great talent, "To build a culture where people feel empowered and valued. I think that is extremely important. Even today, in my new company Embodied (the creator of Moxie robot), I ask how we compete with Tesla, Amazon, and Google in terms of talent. I think it is our mission and company culture, the autonomy that we give our team and the fact that we are bold and not afraid of moving fast forward. And all these things are typically friction points in bigger organizations, like the com-panies I mentioned." He continued to explain that a startup has the added advantage of being nimble without politics and endless meetings vs. a corpo-rate hierarchy and bureaucracy that slows decision-making. In sum, "We [startups] have an advantage because we can build a culture that doesn't have any of those friction points."

Founder Insights: Samantha Lee, Autonomous Vehicles Monitoring

While Pirjanian and Greiner revolutionized home robotics, many analysts pre-dict that autonomous vehicles (AVs) will be the next big thing to disrupt the consumer landscape. Robotaxis and other AVs have hit numerous speed bumps, slowing the prospect of mass adoption. This market reality has led many entre-preneurs to pivot AV tech to aftermarket automobile safety solutions. One example is New York-based Samantha Lee of Meili Technologies, aiming to sell her driver-assist health monitoring platform for trucking fleets and other com-mercial vehicles.

Background

Lee grew up in rural Florida, about 45 minutes from Cape Canaveral, where her father worked on the launch codes for the space shuttle program. This upbring-ing, and explicitly witnessing how her father's epilepsy made him dependent on others for transportation, shaped her product vision. As she described, "My dad lost his license. Probably around the time I was 15, because of his epilepsy, his seizures started to become more frequent. And so the rule is that if you have a seizure, you aren't allowed to drive for six months, and he started having seizures once a week. And so we used to talk a lot about technology and what it could do to help him regain his freedom or in the future what things like autonomous vehicles could do for society." After a series of academic pursuits, Lee set out to create in-cabin monitoring systems for fully autonomous cars but had to pivot when the market came to a halt after many false starts. As she recalled, "We've changed a lot, like a lot of pivots in that regard. And that we've also worked with automakers for cars being sold on the road today with level 2 autonomy, features like adaptive cruise control and lateral steering." She continued, "Things very much slowed down. So we began working actually in the commercial vehicle space quite aggressively, probably about half a year back, and so we've seen a lot of traction there."

Idea Iteration

Rather than investing huge sums of money in Level 5 autonomy, Lee iterated to find a product market fit with today's trucking fleets by meeting everyone in the industry and listening to their needs. "So we've moved into the commercial space, also providing safety systems there, where we've also found about 70% of commercial drivers have pre-existing [health] conditions, too. So for the health emergency side of things, there's a huge benefit there for not only saving lives, because when those trucks crash they cause a lot of damage, but also helping the businesses in that space," explained Lee. Her optimism for an after-market solution that protects drivers with health issues is refreshing after so many high-profile autonomous vehicle startups shuttered. These closures come at a time when the National Safety Council reports that significant truck fatalities have increased close to 50% in the last ten years, a promise that autonomy aimed to solve.

Today's Product Plan

As of January 2024, Lee was working on acquiring the training data to detect people with medical conditions and episodic events while driving, such as erratic breathing, collapsed states, and disorientation. "We're actually doing sponsored research with leading hospitals in New York for heart attacks, seizures, and diabetic emergencies," commented Lee. She then continued to outline her proprietary data collection, "We have about 20 TB of normal driving data already. We're collecting more all the time with our test vehicle, but we're also in the hospital space and actually collecting driving behavior, as well as health events that occur while they're using our vehicle simulator. It's a very niche space where that data doesn't really exist. At this point, we're really the first to collect in mass. You kind of have to go through the hospital in order to do it in a safe way." Lee is not waiting for her training data to be complete before going to market with a smaller version of the platform. "We've done many pilots both in automotive, as well as, the commercial space. The one we can talk about, as most of them are under NDAs, was our pilot with Stellantis. That was the demonstration last year at CES, where we built our system with the 'collapsed state' understanding of incapacitated drivers in one of their Chrysler Pacificas. And so we were showing the kind of responses we trigger like turning on hazard lights, having the vehicle come to a stop, and calling first responders. And what kind of data we would send to first responders," remarked Lee.

In explaining Meili's current sales strategy, Lee clarifies how she is working on driving revenues in the near term. "We actually are selling off-the-shelf hardware components. We would like to eventually be software only, but for now, we're doing this for go-to-market. And it includes things like understanding if somebody has a collapsed state, we have that today, as well as an understanding of those more broad safety systems like limbs outside of the vehicle and backing up incidents [for forklifts]," said Lee. She presented the initial markets for these solutions, "So we began working actually in the commercial vehicle space quite aggressively, probably about half a year back, and we've seen a lot of traction there. We're now a general operator focused on safety computer vision systems. We've also recently moved into the factory manufacturing distribution space." Lee's early traction in enabling uncrewed vehicles, like forklifts and scissor lifts, with vision safety systems is an example of the growing autonomy industry outside of the car.

Lee further predicted that sensor-based technology will become ubiquitous across the industry: "Technologies that enable driving could make us safer today, and even in a Level 5 world." She further predicts, "If you have an autonomous taxi one day, you need to know if people are actually in it, or you could have a big safety issue with other people getting in the vehicle if somebody else is there and they shouldn't be. Comfort is a big focus right now in passenger vehicles already on the road. But that's going to be, of course, even more important down the road with autonomous vehicles, making sure people have more relaxing environments." She reckons this will lead the way for a more widespread autonomous vehicle market over the next 10-15 years. "I'd love it to be on the road like tomorrow, and I'm optimistic. I just think public adoption takes a long time, and that's going to be decades," suggested Lee.

🖆 *Assignment*: Meili's technology crosses many verticals – healthcare, autonomous driving, and shipping/logistics. What team would you assemble to fast-track R&D (training data acquisition) and generate market interest? Be sure to outline the internal processes you recommend installing to ensure a robust feedback loop between users and engineers. Finally, how would you balance project work (like computer vision for forklifts and scissor lifts) to keep the teams with their primary mission of in-cabin health detection?

Notes

1 Sherif, Ahmed . 2023. "Robotic Vacuum Cleaners: North American Market Share 2017-2018." Statista. August 3, 2023. https://www.statista.com/statistics/934290/north-america-robotic-vacuum-cleaner-revenue-share/.
2 NBCUniversal News Group. (2006, May 23). *Soldiers bond with Battlefield Robots*. NBCNews.com. https://www.nbcnews.com/id/wbna12939612.
3 Wikipedia Contributors. 2019. "Dot-Com Bubble." Wikipedia. Wikimedia Foundation. March 21, 2019. https://en.wikipedia.org/wiki/Dot-com_bubble.

3

THE FIVE RULES FOR STARTING AN AUTOMATION COMPANY

Gear

Steve Jobs, a college dropout, tenaciously worked to take the concept of personal computing out of the backroom halls of hobbyists and into the living rooms of every American household. This journey is indicative of a carefully laid out plan that understood not just the lure of technology, but the great marketability of consumer goods. As robot founders yourselves, your journey starts with a blank canvas that we will populate with market data and selling strategies. As Steve Blank likes to drill home, the first step is "getting out of the building" and engaging with prospective users. This will enable you to acquire valuable feedback and iterate or pivot your idea into a novel solution to tackle real-world problems. Using the principles of a lean startup mentality as Mountz shared previously, you will learn how to balance speed and agility to accept failure, as great startups often fail before they succeed.

A valuable resource is to organize all your invaluable market data, gained through your customer discovery, and input "Business Model Canvas" (BMC), which will provide visualization and structure to launching your startup (available widely on the Internet). Using the principles discussed in the first two chapters, you can start to populate your own BMC with product/market relationships, highlighting the different influencers and segmentations for commercializing your invention. Understanding who you are selling to is critical to implementing a thoughtful business model, as well as targeting a sizable market opportunity that could potentially yield millions (if not billions) of dollars. Fellow VC and teacher Steve Blank explains it best; the entrepreneur is on a mission emboldened by his/her "Customer Development Manifesto," updated repeatedly with new customer feedback.

DOI: 10.1201/9781003508410-4

Battlefield Lessons: Plus One Robotics and Eclipse Ventures

High Robot Utilization = Value Proposition

Erik Nieves, cofounder of Plus One Robotics, a leading startup (and my portfolio company) in e-commerce logistics shared with me his guiding rule for all machines – high utilization equals dollars. As he explained, "The first thing you have to qualify, every time, is, do you have utilization. How many hours a day will the robot run?"

Nieves, a 20+ year veteran with Yaskawa Motoman (industrial robots), laid out the key factors to consider for purchasers, "Robots do one of two things. They're either going to move stuff left to right or they're going to perform some process. In the world of just moving stuff left to right, robotics is an efficiency play. But in performing a process, robots must add value to the part [being painted, welded, assembled, etc.] and prove that it is worth more when it leaves the robot than when it came in." This simple guidance answers so many questions for startups regarding value proposition, payback, and even their business model.

"So it became pretty evident early on that process robots were more valuable and easier to justify than handling robots, because of the cost of labor versus the cost of robots. So most of the robots that we deployed in the first 15 years of my tenure at Yaskawa were all process robots, and most of them were welding. The number of robots that were deployed for pure material handling was not that many. When we did it was because the parts were too heavy for people to handle," recollected Nieves.

This insight into historical use cases is enlightening for today's entrepreneurs, as this outcome builds on the first principle of robot deployments – utilization. For example, Plus One works with e-commerce fulfillment centers and Nieves gets asked all the time if he could build something to stack pallets, as this tedious (heavy) task is dreaded by every worker. However, he found that in many instances such a robot would only work two hours a day and be collecting dust the remaining time. The economics of such a convenience does not make sense even for such an onerous task. He declared, "The ROI [return on investment] won't be there with that lower utilization, and that's been true as a first principle from the beginning because the one thing robots were never designed or intended to do was sit still."

Reliability = Customer Retention

In listening to Erik Nieves' advice about high utilization and efficiency, I recalled my conversation with successful venture capitalist Lior Susan, founder and managing partner of Eclipse Ventures, who said it even more succinctly: "Shit cannot fail." He decried the notion of "Fake until you make it," because the former Israeli special forces officer knows firsthand the risks of failure. "People will lose their lives, factory lines will go down," proclaimed Susan, "moving fast and breaking things has real consequences in the physical world" (Figure 3.1).

He explained that unlike software disruption where buyers are more forgiving of network failures due to the infancy of the technology, physical environments by contrast have been around since the beginning of mankind. For example, think back to the first organized militaries of Mesopotamia and the cities of palaces of Assyria or pyramids of Egypt. This is why Susan suggested that these rigid industries (e.g., construction, infrastructure, transportation, and defense) built over thousands of years of human history do not have a stomach for computer glitches and robot failures. It needs to have a 99.99% uptime performance, "Otherwise, they will not use your technology," stated the VC. He continued to underline what he looks for as a backer of uncrewed systems. He further asserted, "And that's the core first principle that I'm a huge believer in – technology must work in physical industries."

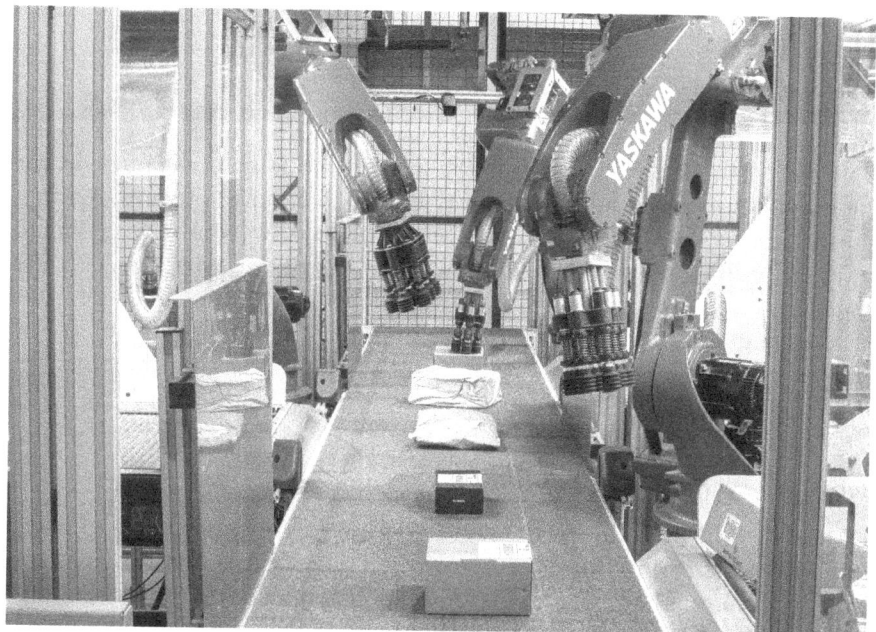

FIGURE 3.1   Plus One robots loading parcels onto a sorting belt.
Photo: Plus One Robotics, 2023.

As Susan talked, I recalled a recent conversation with a collaborative robotics software startup that lost a substantial amount of revenue due to cabling breakage. Hearing this distressing news, I asked the CTO, "How many engineers does he have on staff testing the robot?" He sheepishly replied, "None." I pressed, "How many robots are going 24/7 to substantiate the cycle factor?" Again, "None." This interchange has led to substantial changes in the onboarding and internal workflows at the company with a new department for customer success and quality assurance. The founders have now forecasted their sales to climb, retaining existing accounts and growing wide within these organizations.

Pay Back Factor = Revenues

"The second principle in my book is that technology must pay for itself in physical industries," raised Susan, my new mentor. He continued to share his own experiences as the former head of incubation of Flextronics, "Those people are not willing to buy stuff because it's cool. This is not the way it works in those industries. You need to show them how it's going to improve their operation, lower the amount of people that they will need or generate new types of data they never had. It must pay for itself."

Differentiation = Competitive Killer

In the list of his first principles, the Silicon Valley investor rattled off another, "I think the third one is the moat of building companies, as the intersection of beats and atoms is way harder than pure software. But that means, it's very hard for the second person to do it. So, you're not going to have 30 or 40 companies doing the same things in these markets. You are going to have few, and usually there will be very large companies. "In reflecting on his four principles for a successful robotic company, he stressed again the most important criteria, "ultimately the thing cannot fail."

Target & Low Hanging Fruit

D'Andrea said more bluntly, "You know there are easier ways to make money than creating an automation company. Kiva was a low-hanging fruit that no one had picked, just like the Roomba was kind of the low-hanging fruit for home robotics." He continued to share his latest startup, "What we're doing at Verity with using drones to collect data and insights is another low-hanging fruit. But it's hard. You need super, highly reliable

FIGURE 3.2   Lior Susan briefing Sir James Dyson on his fund's thesis at the Eclipse San Francisco office.

Photo credit: Eclipse Ventures, 2016.

systems that work just like electricity in order to build a whole business case on top of it. And if you don't do that right from the beginning, you're going to get to a spot where you're spending all of your money servicing the stuff that's out there. You're not making any money, and you're just burning through cash, kind of like Webvan did." As one of the three founders of Webvan-influenced Kiva Systems, D'Andrea's warning rings all too loudly (Figure 3.2).

Questions To Ask Yourself:

1 Does my product have **high utilization**?
2 Is my system **reliable, 99.99%** of the time?
3 What is the **payback period** for customers, a year?
4 Have I built a moat around my **intellectual property**?
5 Are my **target customers** low-hanging fruit?

Founder Insights: Dor Skuler and Tom Stevens, AI Healthcare Companions

D'Andrea suggests targeting low-hanging fruit when going to market with robotic innovations, following the five rules laid out by Nieves and Susan. Many governments, healthcare networks, and technologists are scrambling today to address the growing needs of elderly populations aging alone in their own homes. Below is an overview of two competing nascent products on the market that aim to improve senior quality of life.

Background

According to recent census data, by 2050, in the United States, over one-fifth of the population will be sixty-five years or older.[1] This is a massive problem in first-world countries with fewer marriages, older weddings, and children being born. Today, almost a quarter of Japan's population is over 65, with robots deployed in a variety of jobs aiding people with disabilities, including feeding, physical therapy, diet, and exercise coaches.[2] As a result, in 2009, the Japanese animatronic robot Paro received regulatory approval as a Class-2 medical device in the United States. The Wall Street Journal then reported using the anima-tronic device in nursing homes. As Marlene Dean of Pittsburgh Vincentian Homes explained, "Some of our residents need more than we as human beings can provide. We've tried soft teddy bears that talk and move. But they don't have the same effect." This is part of a new trend to use machines to increase socialization in geriatric populations that suffer from cognitive decline and loneliness.

ElliQ: AI Care Companions

In August 2023, Intuition Robotics announced it scored an additional financing of $25 million for expanding its "AI care companions" to all senior households. While its core product, ElliQ, does not move, its engagement offers the first glimpse of the benefits of social robots at mass. In speaking about the future, I interviewed its founder/CEO, Dor Skuler.

He shared with me his vision: "At this time, we don't have plans to add legs or wheels to ElliQ, but we are always looking to add new activities or conversa-tional features that can benefit the users. Our goal is to continue getting ElliQ into as many homes as possible to spread its benefits to even more older adults. We plan to create more partnerships with governments and aging agencies and are developing more partnerships within the healthcare industry. With this new funding, we will capitalize on our strong pipeline and fund the growth of our go-to-market activities."

Skuler further outlined, "We placed very high importance on the design of ElliQ to make it as easy as possible to use. We also knew we older adults needed technology that celebrated them and the aging process rather than focusing on disabilities and what they may no longer be able to do by themselves" At the same time, the product underwent a rigorous testing and development stage that put its customer at the center of the process. "We designed ElliQ with the goal of helping seniors who are aging in place at home combat loneliness and social isolation. This group of seniors who participated in the development and beta testing helped us to shape and improve ElliQ, ensuring it had the right personality, character, mannerisms, and other modalities of interaction (like movement, conversation design, LEDs, and on-screen visuals) to form meaningful bonds with real people," ElliQ's CEO commented. In the testing with hundreds of seniors, he further observed, "We've witnessed older adults forming an actual relationship with ElliQ, closer to how one would see a roommate rather than a smart appliance."

Since deploying in homes throughout New York, the results have been astounding in keeping older populations more socially and mentally engaged. As ElliQ's creator elaborated, "In May 2022, we announced a partnership with the New York State Office for the Aging to bring 800+ ElliQ units to seniors across New York State at no cost to the end users. Just a few weeks ago this year, we announced a renewal of that partnership and the amazing results we've seen so far, including a 95% reduction in loneliness and a great improvement in well-being among older adults using the platform. ElliQ users throughout New York have consistently demonstrated exceptionally high levels of engagement over time, interacting with their ElliQ over 30 times daily, six days a week. More than 75% of these interactions are related to improving older adults' social, physical, and mental well-being."

TomBot: Robotic Pets

According to numerous research studies, animals (especially dogs) have improved moods and increased social interaction with residents, especially dementia patients. Other benefits include decreased anxiety, positive emotions, lower blood pressure, calming effects, and reduced aggressive behaviors. However, pets require a lot of care, and with almost 20% of the population allergic to cats and dogs, having a furry friend permanently in a medical facility is impractical.

In December 2022, I interviewed Tom Stevens, the inventor of Tombot, the most realistic robotics dog made explicitly for treating dementia. He invented the mechatronic canine from the trauma of having to take away his Mom's dog as she had dementia. "I started looking around for substitutes for live animal companions but didn't find anything she liked or would respond to. So I started wondering if technology might play a role that launched me on a multi-year research and education journey which culminated in a Master's degree from Stanford," Stevens expressed. Tombot's latest product, a golden retriever named Jennie, has been shown to reduce stress and violent behavior. Stevens explains that seniors with dementia suffer from emotional detachment as their conditions worsen with loneliness, frustration, anxiety, hallucinations, and, in Stevens' mother's case, violent anger. To date, the medical establishment has relied on harmful psychotropic and opioid medications to ease the condition. However, Stevens' approach is about using the patient's oxytocin production to treat these mental health illnesses by stimulating the emotional support of a dog. Based in Los Angeles, he sought out Hollywood's award-winning special effects studios to provide realism to Jennie. "And then, once that emotional attachment is in place, we become the ideal platform for monitoring that senior for safety and health purposes," shared TomBot's creator.

While Jennie looks like a plush stuffed animal, she's packed with enormous technology. "We did a lot of research and understanding about what it is about dogs that people notice and what makes them attractive. Unlike Spot [a real dog], Jennie is a lap dog with its feet immobile. One of the first things we learned is that anything on the ground presents an extreme tripping hazard for a senior with dementia, and a fall event for a senior with dementia is frequently an end-of-life injury," Stevens affirmed. He suggested considering a robot from the lap dog perspective and the opportunities to be so close to the patient with many sensors monitoring their every move. "So, we covered her with sensors, and she can feel how and where she's being touched. You can tell the difference between a simple touch, a slow caress, a vigorous pet, and being held. She responds to voice commands," he presented. In addition to treating the patient by increasing their emotional attachment, medical professionals can also use the same platform for monitoring changes in behavior, diet, and living conditions. He elaborated on his product roadmap for using this monitoring information to monitor Sundowning syndrome in dementia patients. This condition is where seniors start to be confused or disoriented at the end of the day; to date, it's poorly understood. Tombot is currently working with the Cleveland Clinic to design a doctors' user interface to analyze patient data. He exclaimed that after more efficacy tests, his device would be "The first of any kind approved by the FDA for oxytocin. And with that, we will get Medicare and Medicaid reimbursements."

Assignment: Using the five rules of a successful automation startup, create a scorecard for each invention tackling aging in place. Be sure to analyze all facets of the valuation proposition to drive high utilization, reliability, and quickest payback for its users. Sitting in a VC's chair, which one will have the deepest moat and fastest adoption to become a billion-dollar unicorn?

Notes

1 Vespa, Jonathan, Lauren Medina, and David Armstrong. 2020. "Demographic Turning Points for the United States: Population Projections for 2020 to 2060 Population Estimates and Projections Current Population Reports." Census. gov. U.S. Census Bureau. https://www.census.gov/content/dam/Census/library/publications/2020/demo/p25-1144.pdf.
2 Edmond, Charlotte, and Madeleine North. 2023. "Japan's Ageing Population: The Implications for Its Economy." World Economic Forum. September 28, 2023. https://www.weforum.org/stories/2023/09/elderly-oldest-population-world-japan/.

4

READY TO SELL

Hiking

There is a funny video circulating the Internet with Steve Case, founder of America Online, confronting the admissions officer of Harvard Business School about his failed 1980 application.[1] As Case informs the reader in his book "The Third Wave," he was rejected from every business school to which he applied. In his application's personal statement, he predicted everything the Internet would become now and how it would "significantly alter our way of life."

The vignette provides valuable insight into Case as the entrepreneur. Too often, the last chapter of success is the one that people focus on, however, in this book, we will focus on the journey that includes several missteps and pivots. The important part of your journey is testing your hypothesis and business models on real customers. Now that you have pilot testers, what are you going to do to grow your revenues? Formulating a marketing sales plan to reach more new users and expand your relevance to existing ones will be critical to your startup's success. We will break down these tactics piece by piece to provide a clear launch strategy.

First, Right the Ship

"I think robotics in general doesn't have the luxury of building a nonprofitable business, and I spend a lot of time thinking about profitability," said the former fisherman turned robot investor, Abe Murray, of AlleyCorp. Murray grew up in New England and at an early age dropped out of high school to help run his parents' seafood business. This hands-on experience paid off, and he later earned degrees from the University of Rhode Island, Worcester Polytechnic Institute, and Harvard Business School, ending up on Google's

DOI: 10.1201/9781003508410-5

product team. "I was taught all about that [capital expenditures] by my parents, including that you should fire customers who are not profitable. It's a business fundamental that I see as critically important in the robotics and hard tech industry," described the former fisherman.

"It's funny because I was very good at running P&Ls as a kid. When I was working in defense, our projects needed to be delivered on budget and on time. And then I went to Google. I worried I would lose that muscle of knowing how to run a cost-effective team," reflected Murray. Honing his craft at Google, meant he had to maintain his expense discipline in a place where money was free. "I took my ownership mentality to what I did at Google, and I always wanted to be on teams that were making money for the company and that were doing something really important," shared Murray. He quickly became recognized as a producer for building Google's eBook business and having executive meetings with Larry Page and Sergey Brin about building new businesses. "I had to go in and pitch Larry at a product review and get permission to spend X number of headcount to build a business," the former Googler recounted.

I pressed Murray for guidance for today's mechatronic entrepreneurs: "My advice [for driving sales] is something Google's pretty darn bad at, listening to customers. I was literally not allowed to talk to customers at Google in many ways because of policies. And you can imagine being a product leader at a startup and saying you can't talk to customers unless you jump through some hoops and get somebody else to do it for you. You can't imagine you'd be very successful."

He confessed, "I did it anyways, kind of by hook or by crook at Google. When I brought Google product managers over, I had to push them to go and spend time with customers. Just go and show up." Murray recalled a recent example, "I've met a phenomenal robotics team who wanted to do something in elder care and knew they didn't know the market. And these two young founders literally went and got a room in a nursing home. They lived there for six months and just paid attention; nobody does that at Google. But people do that at startups all the time."

The Tenacity of CivRobotics

One of the best examples of empathizing with your customer's misery is Tom Yeshurun of CivRobotics. Coincidentally, Murray joins me on the Board of Directors of the autonomous mobility startup. "In a nutshell, it was really easy for me to understand the [customer's] pain because I had the pain," said Yeshurun, a former construction manager. The origins story of CivRobotics is one of hands-on (in the dirt) research and product pivots. Initially, the company launched as CivDrone, founded in Haifa, Israel, but shifted to the ground with fleets of terrestrial vehicles.

"When I started working in construction. I was shocked at how old-fashioned things were. I was walking around with pieces of paper, writing emails, and using Excel. And I'm like, there should be a better way," related Yeshurun. This led him to start to experiment with different workflow platforms, but it was not until a costly mistake was made by a vendor that he upped his game. He further shared, "Then I start using drones to capture data and I'm like, 'Wow, this is amazing!'" He spent most of his time overseeing surveying, as his company outlaid millions of dollars hiring consultants to mark and layout construction sites.

"One time the surveyor made a big mistake that was really, really annoying. To be honest, it ended up costing us another 100 grand, which was covered in the end, but a pain in the butt," he recalled. This mistake led Yeshurun to have an ah-ha moment. "I wish I had a machine that would do it for me instead of sending an email to a surveyor. I could just select things on a map and click draw this for me," he declared. This became the impetus for his startup in 2016, as he disclosed, "I knew there was a market for it because I was the customer!"

With his co-founder, Liav Muler, the construction entrepreneur began ideating his concept at the Technion DRIVE Accelerator of his University in Haifa, Israel. This is when I first met Yeshurun on his inaugural startup fundraising trip to New York City. I quickly introduced him to GENIUS NY, an accelerator in Upstate New York for Uncrewed Aerial Vehicles (UAVs). As my firm supported the company with its first capital raise, we began to see there was a problem, a big problem.

As he recalled, "I would say one of the flaws that I did was that I based everything on my love of the drone that we kept on investing in it." In Yeshurun's defense, he had numerous conversations with potential customers who declared, "I love drones; drones are amazing." In reality, he sadly remembers, "But when it came down to it, once we had something physical to show them, they all took three steps back."

Shocked but not discouraged, CivDrones' founders realized they did not have a product market fit with a drone and had to pivot fast. They returned to their prospective customers and listened harder to the feedback on their drone that took them 18 months to build. Surprisingly, two potential buyers expressed interest in purchasing a ground vehicle. Yeshurun and Muller regrouped with the team back in Israel, and then COVID hit. As they now had more time to iterate, Civ started working on version 2.0, funding the development with Israel's Innovation Authority grants for disinfecting public areas. Testing its new autonomous vehicles in malls with Lysol spray gun attachments, the team now had a working prototype. Once the world opened up again Yehurun returned to America and as he bragged, "We did a roadshow, and 9 out of 10 preferred the ground robot prototype."

Then, Roll Out and Prove Value

Everything was now in place for CivDrone to morph into CivRobotics and begin selling, except they still needed a proof of concept. One of the effects of a global pandemic was a work shortage with a backup of jobs. This led to a shortage of surveyors in the solar industry, delaying construction and increasing costs, which is essentially the best-proving ground for a new disruptive technology (occasionally, a little luck plays a big part).

"One of the first customers we worked with had over 800,000 points that had to be marked out. Before [CivRobotics] that was done by 16 people, with us it was four people first and then went down to two people," shared Yeshurun. He continued to paint the picture, "It was painful because it was in a desert terrain, 120 Fahrenheit, and a very annoying work environment with 16 people complaining. Then we gave them a machine, and they literally dramatically cut their costs. We stopped the pain." This led to a simple sales formula, as he stated: "Where the pain is greater, that's where there are more opportunities. And where the labor shortage is higher. This is also one of the reasons why solar jobs have kept on rising. It became more obvious that we should invest more time and effort there."

The result in CivRobotics CEOs' words, "Because of this [market] - easier sale, longer duration, longer volume." As they were the first machine on the job site, technology was even easier in solar, with obstacle avoidance being a non-issue. In addition, the company's sales took off because of the Federal incentives included in the Infrastructure Bill. As Yeshurun said, "There's also a lot of growth in this market, which makes our offering even greater when customers start using us. And then they're going to grow with us." As I write this chapter, Civ's sales are climbing, and requests are flying in from users demanding more robots on the job sites to do more tasks, proving a pivot sometimes can lead to a billion-dollar opportunity.

Remember to be Closing, No Free Pilots, but Lots of KPIs

One of the biggest cautionary mishaps is observing how many startups give away their wares for free in the hopes of earning purchase orders while burning valuable cash. Yeshurun demanded, "There's no such thing as free pilots." He further proved his point: "All the free pilots that we have done in the early days. None of them converted, not even one." Instead, he advises other founders to use money as a litmus test of how serious the customer is about purchasing. He further suggested, "It doesn't matter how much, it could be anything from a grand to ten grand." More importantly, he stated, "They're physically investing capital and their time in this trial period."

Like Mick Mountz, Yesuhurun said that any pilot should be attached to Key Performance Indicators (KPIs). Analogous to Kiva, CivRobotics would turn these objectives to close the sale by asking, "If we meet these KPIs, how many machines do you need?" The consummate salesman, Yeshurun, would also guarantee his products with the statement, "Until we don't make those KPIs work, you can keep the robot at your shop, and we'll keep on improving it." This sales philosophy fosters confidence, and a company seldom loses a customer when there is trust. CivRobotics' first enterprise account went on like this for two years before it would turn into a healthy monthly recurring revenue stream. As the founder recalled, "We went to them once and they gave us feedback. They paid for that pilot and returned two weeks later with the changes. Two weeks later we came again, and the KPIs met. The robot stayed and they ordered more machines. Now we use that case study."

Tom Yeshurun's Sales Wisdom:

- **Try until you buy:** "You want to try us out for a month, here. Rent it for a month. If you like it, great, keep it. If you don't like it; we'll take it back. All good, we'll come back to you later, when it meets all your requirements."
- **Listen to your customers:** "We listen to our customers. Customers love that. When a customer asks for something, we deliver it. They really appreciate it. They're willing to work with us even more. And the fact that we are very customer-driven differentiates us from many other companies."
- **Question adding new features:** "Did anyone ask for it? Is there a reason to invest in development? We're building features based on our customers working with us, saying 'I think it would be a good idea to add this.'"
- **Reliability first:** "As long as the core works well, then you can add more and make some customizations for customers so they're happy."

Founder Insights: Jerry Sgobbo, Underwater Drones

Often, analysts label robots as frontier technology, and there is no place more remote on this planet than the ocean floor. Deep sea drones have the power to provide new information on previously undiscovered realms, but to date, they have been deployed mainly by the military. Mastering the sales formula of Civ Robotics, evaluate how this underwater startup plans to cross the chasm of defense grants to private-sector commercial opportunities.

Background

According to the National Oceanic and Atmospheric Administration, as much as 95% of the oceans and 99% of the ocean floor has yet to be explored.[2] As more than 70% of the planet is covered by water, the promise that uncrewed systems will go deeper into the depths of the sea could be one of the ripest opportunities for autonomy.[3] Already, the Navy has an Unmanned Undersea Vehicle Squadron led by Commander Scott Smith. In an interview in 2019, he explained the military's rationale, "Those missions that are too dangerous to put men on, or those missions that are too mundane and routine, but important – like monitoring – we'll use them [drones] for those missions. I don't think we'll ever replace the manned platform, but we'll certainly augment them to a large degree."

Dive Technologies

Around the same time as the Navy's announcement, I interviewed Jerry Sgobbo of Dive Technologies to comment on how such innovations will proliferate globally and expand into new categories. Sgobbo predicted, "The military and commercial missions have used very similar AUV (Autonomous Underwater Vehicles) technology but are looking for different things in the ocean. Looking forward, both customers are interested in longer-range AUVs. For commercial customers, the goal is to reduce operating costs. For defense, a low-cost, long-range AUV opens new mission sets beyond mine countermeasures and will further lend to keeping sailors safe from dull, dirty, and dangerous missions. Also, AUVs are increasingly important data collection tools for the scientific community."

He optimistically quipped, "With approximately 90% of the world's trade carried across these marine highways, we see the U.S. Navy investing heavily in next-generation AUV technologies to maintain a forward presence and keep shipping lanes secure. As a team, we also look forward to the opportunities we'll discover in the unknown." Sgobbo further described, "We see demand for offshore survey work in the U.S. increasing significantly as grid-scale offshore wind farms are developed over the next decade. In particular, much of this work will take place in New England and mid-Atlantic waters." Sgobbo refers to the recent move by Rhode Island to construct the first-ever wind farm in the United States, capitalizing on the region's famous gale-force gusts. Dive's flexible platform readily lends itself to the development of offshore wind turbines.

Sgobbo further explained, "Dive's AUV is a large platform with very long range and is intended to operate independently without the need for the infrastructure that traditionally supports an AUV mission today. This allows a survey operator to reduce cost as well as perform survey work at times of the year when it is impractical to use a towed system or smaller AUV." The founder added, "For commercial customers, this data is necessary to support deepwater energy infrastructure projects. For defense customers, the same imaging approach is used to locate sea mines."

The startup leveraged its extensive industry knowledge to reinvent how to utilize marine drones. "When we started Dive Technologies, my co-founders and I first took an in-depth look at how medium and large-sized AUVs are being operated and manufactured across the industry today, and we saw vast potential for innovation and improvement," recalled Sgobbo. He continued, "Our new AUV platform, the 'DIVE-LD,' addresses the industry's needs by drastically increasing payload capacity and on-board energy storage but, most importantly, driving down the cost to collect offshore data. We do this by offering quickly configurable payload space to accommodate specific sensors needed for a job or mission, and then letting our robot do what robots are meant to do, operate autonomously and with minimal human intervention." This feature means that Dive's ability to tailor its product to specific mission requirements and greater battery capacity enables it to travel farther and deeper than its competitors. "Today's offshore AUV missions are typically conducted with a dozen humans in an expensive surface support vessel, which leads to important survey work being prohibitively expensive. Dive's novel engineering solution will categorically shift this paradigm," expounded Sgobbo.

Since my interview in 2019, defense technology company Anduril acquired Dive Technologies in February 2022. According to the press release about the merger, use cases for Dive Technologies include: "a variety of defense and commercial mission types such as long-range oceanographic sensing, undersea battlespace awareness, mine countermeasures, anti-submarine warfare, seabed mapping, and infrastructure health monitoring."

Assignment: After reading through the overview about Dive Technologies and its subsequent acquisition by Anduril, how would you coach the sales teams to turn pilots into recurring revenue streams? What do you think is the best business model for such hardware equipment - one-time purchases or subscription RaaS (Robot-as-a-Service) fees?

Notes

1 Case, Steve. 2023. "Facebook Live." Facebook.com. 2023. https://www.facebook.com/watch/live/?ref=watch_permalink&v=1210798718944230.
2 "NOAA Ocean Exploration and Research: World Oceans Day." n.d. Ocean explorer.noaa.gov. https://oceanexplorer.noaa.gov/world-oceans-day/reason-1.html.
3 NOAA. 2019. "How Much Water Is in the Ocean?" Noaa.gov. 2019. https://oceanservice.noaa.gov/facts/oceanwater.html.

5

BUSINESS MODEL AND PARTNERS

Climbing

Investment committee decisions in venture capital firms are far from being a standardized process. In reality, there are a lot of competing factors that play into the boardroom before reaching conviction. Any startup entering a vote of partners faces bias from an investor's past experiences and the individuals presenting. Most generalist firms negatively view hardware, labeling it too capital intensive with long and inconsistent sale cycles. A few buzzwords that linger around investor discussions include investing in "software as a vector for hardware" (meaning the differentiating factor has to be bits not bolts) and the preference for "Robot-as-a-Service" or RaaS business models (translation: recurring revenues like software subscriptions). In reality, accomplishing both these objectives is a lot harder in practice than pitching it in a PowerPoint presentation. Too often founders nod their heads and patronize their check writers knowing full well that their customers have a different calculus in mind.

As stated in earlier chapters a general rule of product development is to forget about technology and focus on the best tool to solve the business problem. Mick Mountz found out at Webvan, and then Kiva, a robot solution became the ideal tool to solve the fulfillment center pick-and-pack problem. If he had just focused on the software aspect of e-commerce, Kiva would've joined the fray of competing warehouse applications versus innovating a new paradigm. Similarly, CivRobotics proves how a hardware-centric approach creates scale, as today's parts and sensors are remarkably cheaper than in recent decades. "Software as a vector for hardware," is not a complete "no" by an investor as most mechatronic solutions today are led by code over mechanics. At the same time, such a backer would probably be scared off by

DOI: 10.1201/9781003508410-6

complete end-to-end solutions like eVTOLs (electric Vertical Take-Off and Landing) flying cars, robotaxis, and new rocket propulsion systems. It's essential to pivot based on user feedback, not funding conjecture.

Clearpath Robotics: Putting Customers First

Ryan Gariepy, the co-founder and Chief Technology Officer of Clearpath Robotics recently acquired by Rockwell Automation for over $600 million[1]) advised startups that "Your customers matter more than your investors do," when pursuing a business model. Too often investors prompt founders to position, or rather contort, their products as service models versus hardware equipment. This potentially could kill a sales lead. Gariepy recommended asking, "How is your customer willing to transact? You can always change business models later. But if no one is adopting your technology right now, then you don't get a chance to change those business models later."

Many VCs try to have every hardware company look like a software sub-scription business utilizing a Software-as-a-Service (SaaS) model when assigning a valuation, often ignoring how most manufacturing purchasing managers buy equipment. Clearpath's CTO refuted this approach, as he opined, "If you can start with RaaS [Robot-As-A-Service] that's great. But I would rather take a smaller [valuation] multiple with a company that exists, where sales are not a constant ongoing struggle than a massive multiple." He advised founders, "Especially in the early days of a company where it's important to get product viability of your company as a whole." Gariepy further shared the primary customer concerns as three hurdles that founders need to overcome when selling hardware subscriptions:

1 **Cost of capital:** "The first factor that comes into play is that many [enterprise customers] can finance metal and silicon cheaper than you. So, they'd look at it and say, 'Why do I give you and your bank the extra spread when I've got a lower cost of capital?'"
2 **Servicing:** "The second [factor] comes into play when you're saying 'I'll maintain all your stuff. I'll service all your stuff for you.' They'll turn around and say, 'Well, no, I want you to teach us how to service it, because our systems have to be operational.' or 'We already have partners we trust who will show up the same day to service our equipment' So that benefit may not exactly hold either."
3 **Gross Margins:** "The third benefit of the gross margin comes into play when you're selling Capex [Capital Expense savings] and that shows up as depreciation on their bottom line. That's actually a good thing. Because many, many companies are measured based on gross margins. But if you sell them RaaS, then it impacts their gross margin in a negative way."

FIGURE 5.1 Clearpath's Husky robot at work.

Photo credit: Clearpath Robotics.

In summary, he cautioned that a bird in the hand is worth two in a bush, even at the risk of displeasing investors: "I would rather sell 100 robots capex right now, and find a way to better monetize later than to really struggle through and not sell" (Figure 5.1).

Rodney Brooks: Setting Expectations from the Onset

Venture capitalist Lior Susan echoed Gariepy's sentiment. "We spend a lot of time on this point, and we do not believe that many of our companies should be subscription businesses, because these industries don't operate like that. That's why SpaceX or Tesla is not a subscription business; automotive and space don't operate using that model, and customers don't buy like that." He further called out unsophisticated hardware investors for badly coaching young founders to follow such business models: "Venture capital over the last 20 years became obsessed with this notion of recurrent revenue. So, everything needs to be SaaS. I think you need to align the business model to the industry that you're trying to transform. Some of them may be SaaS and some might be Capex. Some of them might be robotics-as-a-service. Some of them will be per unit revenue. We spend a lot of time with those companies to understand the best business model for the industry they are trying to transform rather than what VCs like to hear."

Hearing Susan and Gariepy, I tested out their assumptions on "the Godfather of robotics," Rodney Brooks, legendary MIT professor and founder of iRobot, Rethink Robotics, and Robust.AI. He shared with me his first three principles that tie the design of the system to the business model and long-term customer success:

1 **Who's paying for it**: "Figure out who your customers are going to be and who will pay for it. You have to find someone who's going to accept the price that you feel you need in order to make the business work, and if you need both those things to go right for a real business." Formulating succinct sales prospecting and pricing strategies does not stop with the customer discovery process. As mentioned in Chapters 1 and 2, it is critical to re-evaluate constantly, especially when it's time to land and expand in the market.

2 **Setting expectations**: "The physical appearance of a robot makes a promise on what it can do." He illustrated with an example of a humanoid designed to look like Albert Einstein, "And you know it's not as smart as Albert Einstein. It's probably going to disappoint the end user. They feel it doesn't know anything." He further demonstrated an example from his own experiences at iRobot, "The Roomba was great, as it didn't make a big promise. It wasn't Rosie from *The Jetsons*. It was just this round thing that rolled around on the floor. It didn't make a great promise, and it delivered. But if it looked like C-3PO, it wouldn't have delivered on it. Think about how your robot looks and what it is promising."
Since speaking with Brooks, I've heard his words about robot physicality linger in my brain as today's humanoid robotics industry is in a venture funding bubble, amplified by Goldman Sachs's latest research predicting android robot sales to reach $35 billion within the next ten years. Could these cyborg startups be setting expectations too high for early adopters to achieve this level of deployment? By contrast, collaborative robots (first installed 15 years ago) have just hit the $1 billion global sales milestone.

Physical appearances ease the fears of working with machines, which is why Brooks at Rethink Robotics designed a friendly head with eyes for his cobots, Baxter and Sawyer. Sharing her perspective Andrea Thomaz of Diligent Robotics, an autonomous collaborative healthcare robot, retorted, "A big part of our technology is its social presence in the hospital. We really want to think about Moxi as being a robot that's part of the team. It's joining the service team of people in a hospital, but that also means we are wandering around in the hallways with patients, nurses, and staff. We heard from people that they really wanted the robots to be a bit more social and interactive in the hallways during transport. We ended up adding a lot of things that were expressly just for the social interaction." As she elaborated concerning Diligent's humanoid-like product features, "So now as Moxi (the robot) sees a person it flashes heart eyes and beeps and those are some of the things that we hear the most about from the hospital staff. I think it almost surprised us how much people appreciate the social agency of the robot in the environment. The former MIT researcher added, "That is why it has a head is to really express social agency and express intentionality so that you know where its head is facing and where it is intending to go, and that is super important for a robot that's going to be in a social environment."

3   **Providing human agency**: Brooks continued, "Robots are not going to be perfect. You can have controlled environments, and there they can do the same thing over and over. But when you enter an uncontrolled environment, they won't be perfect. They're going to fail. If your robots don't give the humans around any agency around their failures, they're going to get pissed off at it." He then shared a catastrophic example of a robot failure, "If it's in a hospital taking dirty sheets down to the laundry and there's all this commotion in front of it with nurses pushing a gurney with a patient about to die running down the corridor. Then it [the robot] stops and blocks the corridor and the nurses have no way of getting out of the way. They hate the robot. The robot will not survive in that environment. You better have agency for the humans around [the robot] to take over, because the people around are always going to be smarter than the robot in this physical situation." Examples like this drove Thomaz to add more human agency elements into Moxi as it scales across hospitals throughout the United States. Brooks then elaborated on how his team infused human agency into Robust.AIs DNA with current product deployments.

The Handlebar Example: "Now our CEO had this idea of why not put a handlebar on a robot, because we were getting worried about robots in hospitals, and how they were pissing people off because they took away people's agency." The epiphany for adding a sensor-loaded handlebar to their Autonomous Mobile Robot (AMR) came from the CEO's vacation with his wife and kids. Anthony Jules told Brooks, "We should put a handlebar on the robot. Then people will just be able to grab it and move it out of the way when it's causing problems. Now, it has become more like a shopping cart for warehouses, with zero automation and terrible labor. Sally Miller [Chief Information Officer] at DHL saw the possibilities immediately." A great example of how value engineering for users leads to closing sales.

ROI = Zero User Friction

Brooks put forth an alternative point of view on subscription business models. Unlike Susan and Gariepy, he said that Robust.AI is selling their mobile robots in the warehousing space on a RaaS basis. The Godfather of robotics shared, "We're trying a new business model that I haven't done before, which is robots-as-a-service, but it's related to what I think is important - have zero friction for adoption." He reflected on the Roomba as the perfect example, "The promise was you take it out of the box. You don't read the instructions. We ended up eventually not putting instructions in there. And you switch it on, and it does its thing. You don't have to do anything. Zero friction means there's a return on investment immediately. The first time you ran it [the Roomba] and it went under the bed filled up with dirt and then beeped, you got an immediate return on your investment." Returning to Robust.AI, he shared, "So that's what we are trying to do with our warehouse robots, there's an immediate return on investment." Brooks stressed that ROI calculations are critical for deploying products, by setting the right business model upfront, a startup can manage user expectations from the onset of the relationship (Figure 5.2).

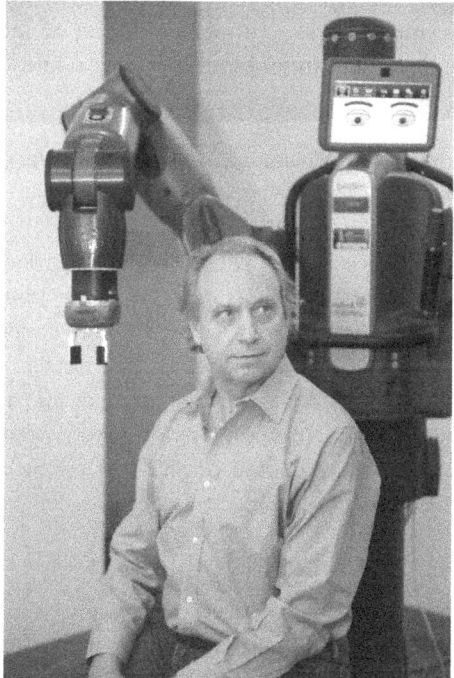

FIGURE 5.2   Rodney Brooks featured with Baxter robot.

Photo credit: Rethink Robotics, 2012.

Brooks' astute observations aren't just for selling consumer and warehousing robots but across the board. As he remarked, even military operators who trained all their professional lives for conflict want zero friction. While many know iRobot as the inventor of the Roomba, Brooks, Angle, and Greiner were also responsible for one of the most successful military uncrewed systems – the PackBot.

The former founder nostalgically recalled, "When the PackBots were first deployed in Iraq and Afghanistan, they had this horrendous menu-based interface. And who were the users? They were mostly 19-year-old kids who graduated high school and joined the military. Now they're getting shot at with roadside bombs. They want to get this robot out there to do something with the bomb. So, we changed it. We made them with Gameboy controllers. Now every operator with zero training has 10,000 hours of experience. We made the controls look like a video game. And now they could do something."

He recollected how his engineers first pushed back on the change, saying "If only the soldiers with bullets flying at them would read the instructions, they would know how to use the interface." Brooks in his wisdom was focused on making the PackBot as easy to use as possible. The result was the soldiers could stay in the Humvee and deactivate a bomb with the robot. These Pak Bot users cheered the update, "I'm in control of the robots without getting out of the vehicle where a sniper could get at me." Brooks stated, "That was their return on investment in the first, you know, 5 minutes." This

Founder Workshop: Atle Timenes, Automating Warehouse Pallets

Brooks and Gariepy are tackling the multi-trillion-dollar online fulfillment, logistics, and manufacturing market opportunity. Like Kiva, labor shortages, and consumer demand have created a massive shift in supply chains and distribution channels. Most inventors that we have discussed are creating new autonomous systems. By contrast, wheel.me aims to modernize existing pallets, rackets, and trolleys into uncrewed systems (basically a smart castor system). Unpacking the description below, grade the proposed original business model to its existing sales strategy (on its current website).

Background

There are 22,000 warehouses in the United States; most analysts estimate that 80% of them have yet to be automated.[2] A big hurdle is upgrading these extensive facilities, which could be three football fields in length or $2 million to

$20 million in robots, depending on the square footage. The lifeline of these fulfillment centers is pallets. The United States alone used more than 3 billion pallets in 2023.[3] The global pallet market reached $66 billion in 2022 and is on track to climb to $110 billion by 2032 with the adoption of more environmentally friendly reusable plastic materials.[4] Given that distribution managers purchase or rent pallets every day, the value proposition of adding inexpensive technology to this existing storage platform could be a game changer in revolutionizing the industry.

Reinventing the Wheel

The idea of reinventing a 5,000-year-old product was first proposed to me in 2019 by Atle Timenes, the founder of wheel.me. As he explained the idea's origins, "Our journey started with the inventor Rolf Libakken (current wheel.me product developer) helping his family move several times during a short period. The obvious question when moving heavy objects was, 'Why aren't these things on wheels?' A mechanical 'click-wheel' was invented, much like the principles of a ballpoint pen, also serving as wheel.me's initial commercial product." This click-wheel caster was introduced in 2013, which added manual rolling motion to any static object. Fast forward a few years: Timenes and his team upgraded their patented solution by stuffing it with a motor, gearing, batteries, and vision sensors into their cylindrical sleeve [today it is more like a box]. He further outlined its complete command and control platform, "Positioning and navigation data are sent and received from the cloud. The system can run on Wi-Fi or a local network. Data science capabilities are added for a back-end/cloud solution and with swarm-enabling technology wheel.me can perform coordinated movements with a large number of objects."

The Norway-based entrepreneur expounded, "Automation is rapidly changing the landscape for businesses. Still, early adoption could lead to risks affecting financials, quality, safety, and process. wheel.me takes down entry barriers to adopt automation technologies due to orders of magnitude lower costs and charging according to an indoor mobility-as-a-service model, making pay-back and return on investment highly attractive".

I pressed Timenes on how he plans to compete against larger competitors that use automated guided vehicles (AGVs) and autonomous mobile robots (AMRs), and he argued, "Existing solutions require frequent human intervention to load/offload and manage the object being lifted, dragged or pushed. wheel.me finds it counterintuitive that objects that already are fitted with wheels (or easily could be fitted with wheels) should need a 'device on wheels' to be able to move them." He continued, "In line with the classic description of a disruptive innovation, we believe that current suppliers of automated indoor mobility and materials handling solutions are on a performance trajectory that is tailored to the high-end of the market." He believes in making any object smart with robo-wheels from event chairs to parking systems to manufacturing equipment. Timenes is working on a new startup today, but his optimism is still rolling through the autonomous wheel company, which has raised over $50 million.

Assignment: You are now in the investment committee boardroom presenting wheel.me's concept, business model, and marketing plan to build consensus and conviction for the investment. List the questions you anticipate your partners asking you when challenging the business, and come prepared with answers.

Notes

1 Staff, Robotics 24/7. 2023. "Rockwell Automation Agrees to Acquire Clearpath Robotics, OTTO Motors—Updated." Robotics 24/7. September 14, 2023. https://www.robotics247.com/article/rockwell_automation_acquiring_clearpath_robotics_otto_motors.
2 "Global Warehouse Automation Robots, Technologies, and Solutions Market Report 2021-2030 - ResearchAndMarkets.com." 2021. Www.businesswire.com. June 21, 2021. https://www.businesswire.com/news/home/20210621005532/en/Global-Warehouse-Automation-Robots-Technologies-and-Solutions-Market-Report-2021-2030—ResearchAndMarkets.com.
3 Munholland, Glen. 2023. "Wooden Pallets Cost Us More than You Know!" Circular Supply Chains Inc. March 6, 2023. https://circularsupplychains.com/2023/03/06/wooden-pallets-cost-us-more-than-you-know/.
4 "Pallets Market Size to Hit around USD 124.9 Billion by 2030." 2023. Www.precedenceresearch.com. August 2023. https://www.precedenceresearch.com/pallets-market.

6

CAPITAL PARTNERS

Wildlife

The first thing one learns on the hiking trail is that water is vital to life, without it plants shrivel, and animals (including humans) die. The same is true of cash, capital is the lifeblood of any business. The challenge with hardware is similar to scaling a rock face; perilous costs are associated with launching such a startup, including expensive talent, sensor components, and proprietary data infrastructure. Even the best-validated plans will experience the arduous task of finding the right partner, as the pool of mechatronic investors is small and increasingly cynical with the post-2021 private capital valuation collapse, exacerbated by the dramatic decline of publicly traded reverse mergers (also known as SPACs). The key to navigating the chaotic path is first realizing that not all financial backers are the same, aligning synergies is the first step to embarking on a successful journey. This chapter will introduce you to leaders in venture capital, corporate ventures, government agencies, accelerators, and academia to help navigate the critical next steps of your business.

> Pitching Tip: make sure you have a succinct investment presentation (i.e., the deck). While there is no set format, Venture Capital Fund Sequoia has a very good template (based on Airbnb's original presentation) available on SlideShare.com.

Diligent Robotics: Hard Tech Fundraising Boot Camp

"We were raising for a hardware company, selling into healthcare in 2017. It was brutal as deep tech and healthcare investors were almost zero. The thing that you run into as a hardware founder pitching VCs is you often get people

DOI: 10.1201/9781003508410-7

who push for hardware companies to have recurring revenue - SaaS metrics. So, you have to find the investors that get hardware businesses are going to have different metrics," counseled Diligent Robotics' Thomaz.

To further emphasize Thomaz's point, one must understand that the pool of capital partners is tiny compared to VC-interest software-led companies. In addition, most funds have little appreciation of how equipment is purchased and paid for in the enterprise, as many analysts previously worked as consultants and bankers. To overcome this bias, Thomaz educated the market on the long-term benefits of robots, "Investors that like investing in hardware companies get that it isn't a SaaS company that's going to be able to go from 1 to 10 million, but the churn is going to be really low because once a hospital has invested in installing robots there done with change management."

Finding the right capital partner is critical; as Thomaz advised, "Be super selective of who you decide to raise money from because you're going to be working with them for a very long time, through tough decisions. It's made a really big difference that we have strong partners that really believe in the vision and really support us." I always suggest that just like VCs conduct due diligence on prospective investments, so should founders on their financial backers. This relationship is a business marriage; your startup is the business child. Custody disputes are ugly, so recon is critical before signing away your shares.

Venture Capital Playbook

Too often founders think of venture capitalists as a homogenous population of executives running around with bags of money, like the monopoly man. The truth couldn't be farther from reality. Every firm has its own thesis, culture, and driving principles. For example, my fund (ff Venture Capital) does not invest in life sciences; however, about 20% of the deals sent to us by founders are in healthcare every week. This drives us bananas, as anyone who would've spent five minutes on our website would realize it wasn't a fit. Rather than blanketing the industry with emails, LinkedIn Messages, and WhatsApp texts, my advice to startups is to be very targeted starting with the fund portfolios that align best with the technology you are pitching. Remember everyone is six degrees from someone who knows someone at a fund, so use LinkedIn or other relationship management databases to rustle up warm intros (cold email campaigns and other messages are the easy way out).

Besides the vertical focus of different funds, there are also stages of investment. If you are raising a pre-seed round pitching to a growth equity fund, your effort will fall flat; as stated above, perform your due diligence! Finally, as much as you are pitching them, so is every other startup, and most firms' deal rooms are overflowing with prospective investments, leading to swift decision-making (often in the form of an online thumbs up or thumbs down vote in an Intranet database).

As the Field Guide aims to help founders start a robotics company from the ground up, below is an outline from the investor's perspective of the three main ingredients that most VCs use to evaluate early-stage investment pitches:

People: Unlike late-stage investments, which financial models and sensitivity analyses drive, early-stage conviction starts with a psychological of the team view as investors ask if the founders pitching are the "right people" to build a billion-dollar business. As Thomaz experienced early on, "People are investing in the team and the idea." Questions ranged from, "Do you think there's a huge market?" and more pointedly, "Are you the team that's going to be able to make progress quickly?"

Hearing the investor side, Lior Susan of Eclipse Ventures expressed, "We really like founders that understand, incredibly well, the market that they're trying to transform." It's impossible to replace real-world valida-tion, as too many would-be technical founders focus purely on simulations and concepts before figuring out if there is a void in the market. As Susan advised, "Industries are huge. It's very easy to research. Who are the incumbents? What are the trends in those markets? And if you're building a company to transform one of those industries, you better know what you are talking about. Do a lot of research, and be intimately familiar with the way that you're going to roll out your solution, the ROI, and who are the best customers than others. Unfortunately, we find a lot of people focusing too much on technology at the expense of a very light understanding of the market that they are trying to transform. This for us will be a turnoff."

I don't know how many times during a pitch I find myself googling market data, providing me with more customer information than the founders on the other side of the table. In addition to knowing the market opportunity, a founder must master his/her numbers. If you have to say, let me look this up, you fail. At the same time, one must be careful of tempering their domain expertise to avoid coming off as a know-it-all, as arrogance is probably the biggest turnoff for prospective partners. This is a business union, and inves-tors will be there for you to ride the wave, coach you through tough times, and evangelize your products to the world, but only if you are open and flex-ible to listen to their advice. Passion is, of course, critical to any startup, zealotry on the other hand could be a death spiral. As Abe Murray of AlleyCorp described investors' mentality, "You're a little bit of a therapist, a little bit of a listening session, and they work through the problem with you." He compared his activities as a sounding board to the engineer's rubber duck, "If you're debugging code, you put a rubber duck on your desk and you explain to the rubber duck your problems. Then the rubber duck will help you solve your problems. Engineers do this all the time, talking to an inanimate object to solve problems, we [investors] are like the rubber ducks for our founders." Murray's philosophy is further emphasized by Lior Susan of Eclipse who boasted, "When going to a good VC, you're really getting a partner that's going to help you grow and scale. That's our model."

> **Pitching Advice for Women:** I hate to admit it as a man, but female founders are treated differently by the startup industry and its funders. Institutional bias has even informed twenty-first-century investment decision-making. As Thomaz informed, "In all honesty, it is extremely important for a technical female founder to brag about herself a little more than she's used to when she first walks in the room because most people assume that you do the biz dev and the marketing and there's some tech guy that builds the robots." This advice was given to her early on from a female VC after her first pitch upon learning about the founder's PhD from MIT and work in the robotics lab: "She was like, 'Oh honey, you got to say that first. That needs to be your first slide.' I was like, 'Really? That seems so weird. That's not what they did in the Airbnb deck.'"

Product: Once the founding team has passed the sniff test of humility, integrity, enthusiasm, and authority, it's time to unpack the product. This analysis quite simply comes down to two questions:

1 How mission-critical is the product?
2 How defensible is the intellectual property?

The first question is asked several ways, for example, some categorize technology in terms of features vs. real businesses. Features are just added benefits to existing technologies (like an easier database tagging system), while businesses monetize the added effect of a complete solution (such as a robotic cell for manufacturing). Lior Susan said best, "Go big and bold if you're going to conquer an industry! That makes the most sense and will have a huge impact versus small and incremental."

My partner, John Frankel, would always ask at investor committee meetings if the proposed technology is 10x faster, half the cost, and more accurate than a human worker. If not, he would veto the deal. Too often, roboticists present solutions that are targeted at long-term capex savings, but move at 20% the speed of a human and cost more than double the hourly wages. As stated in early chapters [probably a good time to review], product market fit is determined by high utilization, reliability, and customer payback.

Mission critical means solving real industry problems, not deploying cool tech because it's available. While humanoids and generative AI are sexy today, if you do not have a use case for these exciting new applications that customers are demanding and paying money for, you will languish regardless of the market hype. PT Barnum cried, "There's a sucker born every minute," tying your company's valuation to hyperbolic market trends runs the risk of wasting time, money and creating unrealistic return expectations that astute professional backers will reject (this will be discussed further when reviewing deal terms).

Comprehending the other side's point of view is crucial to presenting a clear value proposition to investors. Founders often fail to appreciate the business of funds and to increase the wealth of their limited partners (high net-worth individuals, family offices, pension funds, endowments, and other institutions). For a firm to reach an investment conviction they have to believe your startup will drive an alpha return over and above the market in the form of a positive liquidity event (secondary sale, private equity buyout, corporate merger/acquisition, or initial public offering). Many funds also have ownership and capital commitment thresholds that high valuations make it impossible to achieve. While having an exit mindset is important, Susan shared that his fund's philosophy is more focused on the value creation of enduring companies, "Building great companies, and when you build great companies, good things will happen, and you will have options."

Founder's Perspective

As Thomaz experienced, one way of launching fast is by literally living at the customer's facilities, leading Diligent Robotics to drive early market adoption, and as a result, today has cumulative investment dollars of over $75 million. "Hardware takes a long time, so one of the first things we did was get out in front of our customers as fast as possible. We raised a little bit of seed capital in 2018. Within nine months, we had a first prototype built, and we were embedding ourselves in hospitals, testing both technical feasibility and product market fit. We figured that out pretty quickly, we're able to bring a Robot-as-a-Service model to a risk-averse market like health care. Having a solution where people could just pay as they go a little bit was really great for getting traction quickly," Moxi's creator shared.

It's not enough to entice capital partners with a product vision, you must also show that it is unique and defensible in the marketplace. Early funders fear that larger established players will replicate the product and crush the startup once the product becomes known in the industry. I mean who would invest in a new search engine and go up against Google? Your technology must be proprietary and protected. As investors see a lot of deal flow, leading quickly to cynicism, many will try to debunk swiftly the technology as "been there," "done that," and "all smoke and mirrors." Note: adding .ai or .io to the end of your domain name does not mean you are now an AI or IoT company.

It is your job to show how there is a market void, and you are filling it with a novel system that can become THE new workflow platform for your industry. There are several ways to protect your ideas through exclusive algorithms, trademarks, patents, and most importantly the market. Sales change everything, and if you are the market leader you have a head start on everyone else. Smart companies also look at how they can dig a moat around their intellectual property. For example, by entangling your technology within the broader enterprise, you can make your customers' switching costs as high and as cumbersome as possible. This entanglement provides added fortification against upstarts and established players underselling you. The last thing anyone wants is a price war and a race to the bottom (Figure 6.1).

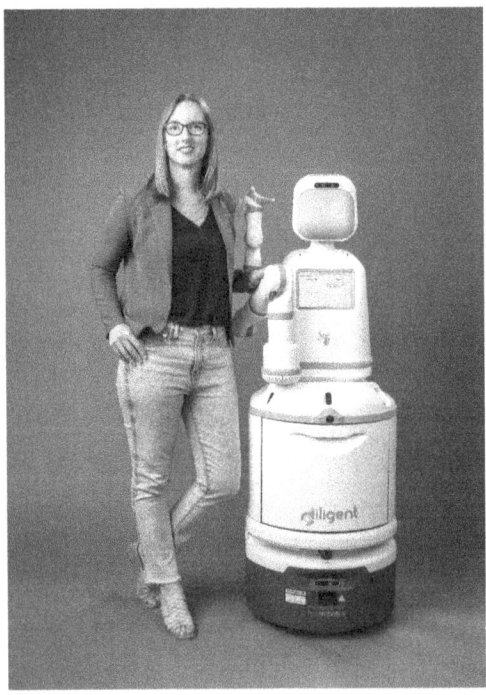

FIGURE 6.1 Andrea Thomaz with Moxi.

Photo credit: Diligent Robotics, 2022.

Cautionary note: I have 10 patents in my name, which is a great way to spend money, but I'm under the belief that startups are still better spent on generating revenues. Being proprietary doesn't mean you have to build everything yourself. As Murray noted, "I'm seeing a lot of new founders who are just assembling off-the-shelf components in interesting ways and running a lot faster." To translate, pick a lane. You don't need to create everything; you just need to build the parts that distinguish your idea from the rest.

Plan: There's a saying among VCs if you want to do due diligence – "customers, customers, customers." While many businesses we see are pre-revenue, sometimes even pre-product, none are pre-validation. In the first chapter, I discussed the importance of customer discovery. Now, it's time to apply the lessons learned by early adopters to impart your vision to financial backers. Remember, VCs desire billion-dollar business potentials. Many automation startups often take on professional service jobs to pay the bills. This income source is a great way to raise inexpensive capital, but most investors look down on consulting projects as an exciting business model. We like platform-led technologies that land and expand over an industry by "rinse and repeat" sales – deploy the solution, solve the problem, and replicate the business. Once in with the customer, it's time to expand the relationship to tap into the full customer lifetime value. The first step is securing a pilot or test site for your machines. "You might hear the term big logos," meaning Fortune 500 corporations as your early adopters. Often these large organizations have innovation departments that scout out new market trends to adopt internally, the problem is these pilots are often far removed from the business units. You must show investors you have a path to commercializing your technology, rather than staying in pilot purgatory. As Mick Mountz suggested, tying your deployments to the KPIs of the organization is critical to moving the ball forward and scoring touchdowns.

Founder's Perspective

In speaking with Charlie Andersen, founder of Augean Robotics (a ffVC portfolio company) and maker of Burro, an autonomous farming robot, about raising capital he exclaimed, "Sales solve all problems. People lose sight of that, especially in robotics. I think ultimately all VCs care about is, 'Can you sell a bunch of stuff?' And if you're selling a bunch of stuff, then the logic is that as you sell more, your costs fall on a per unit basis, which enables you to achieve scale." He recalled, "My early pitches were pretty consistent showing how we were getting into the field early and often, accumulating a ton of customer feedback, and then trying to prove out the sales model as rapidly as you possibly can. And that hasn't actually changed a whole lot throughout time. Team, key milestones, product technology, defensibility, and those things are kind of second to proving out that you can actually sell whether you are raising at seed, A, or B."

Sales come in many shapes and sizes, starting with paid and unpaid pilots to direct sales and licensing arrangements. As Andersen cautioned, "If you're doing a pilot in some form and it's not paid, you're wasting your time." Your pitch must answer whether and how you plan to foster third-party sales, such as through channels, system integrators, dealers, and other partnerships.

"Elephant hunting, not squirrel hunting looking for a big customer to sell to direct," he added. "As we've gotten more and more mature, we're now starting to add in dealers."

Andersen commented that the timing has to be right to add these vendors. "Dealers are a lot like Bob Dylan plugging in and going electric," he said. "If your music is good, then what is amplified is going to be good, but if the music is not good and you plug in the end, what you have is more noise."

Andersen continued to share Burro's international growth strategy of leveraging dealers: "We've evolved and started to sell through dealers. We found that it's really important to segment our product line and only sell the stuff that is very, very mature through dealers rather than selling some of the novel new stuff, which isn't ready for primetime in Australia, New Zealand, Japan, and other parts of the world." The results for this Series B agritech company are evident as the founder shared, "I've got roughly 25 or 30 dealer locations today. It's probably about 40% of my volume. It is definitely evolving in a positive direction, but it took us some pretty big steps to get there, some of which were candidly pivots."

Daniel Theobald of Vecna Robotics Cautions: Read the Fine Print

There is a natural progression for startups first raising funds from the people who know the founders best – friends and families. There are two financial mechanisms for receiving capital, debt and equity. As equity rounds are very costly and involve intricate legal documents, sometimes going on hundreds of pages in length (investor rights, operating provisions, and other shareholder agreements), early founders opt for debt instruments such as a Simple Agreement for Future Equity (SAFE) or other notes that convert to equity upon qualified financing. There are terms to be aware of for founders, such as capped notes (the estimated conversion valuation) or uncapped (postponing any valuation discussion).

Also, often founders provide incentives for these early risk takers such as discounts (typically ranging from 15 to 25%) to the share price of the equity conversion. Once an equity round happens, the lead investor will set the terms. Too often, in my opinion, new founders price their company's value based on the last industry press release. This approach wastes critical time and resources that could be better spent building products and selling them.

Daniel Theobald started Vecna Technologies shortly after graduate school, which later spun out the robotics company. Over time, Vecna Robotics evolved from a professional services business to a product-driven company, always bootstrapping the entire enterprise through revenue contracts. Unlike most founders who rush into venture capital relationships, he prudently waited to bring on investors until his business was at the right inflection point. To date, Vecna has raised over $200 million (Figure 6.2).

Theobald warned, "I feel the mistake that many people make is they are taking money because they feel like that's how you're supposed to do it, and that is not a good reason to take money. They feel if I'm a real company, if I'm a real founder, I'm going to have to convince VCs to give me money. Well, I will tell you that the real founders, the real successful talented people figure out how to do it without using somebody else's money. And I'm not saying that this is the right way to do it, there's no right way to do it. There are a bunch of tools available at your disposal, and choosing the right tools depends on the details of your specific situation."

FIGURE 6.2   Daniel Theobald founder of Vecna Robotics, working on his latest uncrewed venture.

Photo credit: Daniel Theobald, 2024.

Instead, Theobald recommended that the best time to raise money is when things are going well. "If you get to the point where you've proven product market fit, you've got a growing customer base, and you now need to ramp up production, which means you're going to have a huge capital outlay to build a factory," he said. "Well, that would be a great time to think about taking in outside money, because by not having that money you are literally just limiting the growth that the company can experience. It's like starving a growing plant of water."

He further shared the dangers of having too much money, too early, "Clayton Christensen, *The Innovator's Dilemma* [author], was a mentor of mine. I was meeting with him once, and he said, 'After I wrote my first book, I became very famous in the business world. Overnight, investors approached me, basically throwing money at me. So, I decided, I'll start a company. Why not? We had so much money, it was falling off the table,' his words. And then he paused, and he looked at me, and he said, 'That was the worst possible thing for us. Then the reason was that it created an artificial economy. It created a sense that their company was successful. Because boy look at all this money. We must be successful.' They weren't successful. They had no business model that worked. And, you know, eventually, the company just fell apart because they were not forced to kill what they eat." In response to this tale, Theobald advocated, "I think this idea of bootstrapping can be very, very powerful, or at least, going through the exercise of – could we do this? how could we fund this, versus just jumping straight to – I'm going to go get money from somebody?"

As Andersen stressed, "I think what's surprising to me is there are actually a remarkable number of robotics companies that have raised 50 to $100 million in cash and have under a million dollars in sales. If you think about it, that's your worst nightmare. Ultimately what investors care about is can this company build a category that has some significant scalability to the point that the category becomes something that a large other company might want to acquire or be a great standalone business. What founders should be trying to do, in my view, is build a great revenue and a high margin business." Rather than focusing on the dilution, it's more important to find a lead investor who can set the terms of the round for all others to follow and give you breathing space to execute your business.

From GM to Amazon to Yamaha: Corporate Venture Secrets

While venture capitalists exist to make money, corporate venture capital (CVC) concentrates primarily on discovering strategic fits for their business units; at the same time, the investment needs to have promise, as no one likes to look bad and lose money. Sean Simpson, formerly of Yamaha and GM Ventures, explained, "A corporate venture unit at its best is one that has financial returns and great investments that are also strategically relevant. Some of those will end up turning into real engagement. Some of them won't and that's okay. You shouldn't have a hundred-percent hit rate, or you're not taking enough risks."

When founders interface with corporates, it's crucial that they understand the priorities of the internal champions who are excited about your technology. "For a company to have strategic benefits, it needs to be financially viable. It needs to have other investors. It needs to be making progress on its goals and ultimately growing to stay alive and actually add value. But at the same token, you can't just be completely financially motivated, because, at the end of the day, we always said this at GM Ventures, 'If we invest 5 million and we have a 10x return of 50 million, it's not even a rounding error in terms of GM's bottom line.' And so it's really not why we're doing it." "To repeat, the bigger the company, the bigger the rounding error your potential return could be, so it has to be about more than money," Simpson maintains.

To sum up, he says you need both financial performance and strategic fit that enables internal stakeholders at the corporation to trust the technology enough to push it through the bureaucracy. Startups need to note that corporate venture groups seldom lead rounds and to hook a strategic partner, you still need a quality lead VC investor to whom all the other shareholders look for deal terms and due diligence.

How CVC Works

The majority of my portfolio companies, especially in the hard tech space, have CVCs on their cap tables, such as Amazon (Plus One Robotics), Festo (Cambrian Robotics), Trimble (CivRobotics), Toyota (Burro), to name a few. While many investors demur the thought of a strategic partner having such a close relationship with a portfolio company, I believe it depends on the organization. A major concern could indeed be exclusive access to proprietary IP that a corporation could look to acquire on the cheap before becoming a competitive threat, however in reality this would only hurt the CVC's ability to invest in future deals. Most corporations utilize innovation to offset their R&D efforts by adding technology to their supply chain and riding the wave of its value on their balance sheet.

Simpson illustrated the intangible advantages of CVCs through how his work at GM in helping an Israeli IoT cyber security startup, Claroty, further develop its product for the automotive industry. He elaborated, "It seemed relevant to GM in the sense that we had a lot of critical manufacturing assets, but we actually didn't have anyone responsible for protecting those assets. I knew I wanted to find a way to invest in it, and I knew it was relevant to GM. For most CVCs, you need a sponsor [business unit]. You need to have someone within the parent company, not on the venture's team, who is going to benefit from this work with them. Ultimately, they [GM] put a person who was more of an IT cybersecurity expert in charge globally. That's really when we were able to make progress."

He continued to outline how the two companies helped each other improve performance. They weren't working with automotive companies then, but they saw GM as a good representative of automotive companies. If they can solve some nuanced issues for GM, there will likely be a lot of other customers in the automotive ecosystem. Once they [GM] put someone in charge of it and engaged with them, he was able to say, "Here are some of the changes that might need to be made and ultimately were implemented." Claroty designed its product to meet the needs of GM's large manufacturing operation and grew from there. As Simpson noted, "That was able to let us get into the round that had already closed, by justifying the investment with the fact that we were going to really help them develop their product for a new market as their anchor automotive customer." GM engaged with Claroty, which recently raised $100 million at a $2.5 billion valuation, in a variety of ways as a customer, strategic partner, and board advisor that would've not been possible with a typical investor.

Tips for engaging with CVCs:

1 **Don't Rest**: lean into the relationship, don't stop, as early excitement around the financing often doesn't carry forward without personal investment. This means you need to prop up your internal champion, attend internal trade events, and find ways to motivate teams to support your sales. For example, one of my portfolio companies forged a relationship with a leading supplier and reseller of equipment within their space. The CVC group promoted their investment in the startup as the future of the company for digitizing their products and services. One of the sale points told early on to the startup for including the CVC was help on the supply chain regarding pricing, order quantities, and delivery times. This did not translate to the business unit responsible for selling the goods, which quoted higher prices and longer lead times to my startup. The startup has persevered by attending internal sales gatherings, sharing R&D feedback, issuing joint press releases, and creating cross-selling dealer incentive plans.

    As Simpson suggested, "If you're selling a robot, you're selling real atoms. I think it's super helpful to be there in person. Go check out their innovation facility. Let's say they have 30 projects and half of them are covered in dust, which tells you something. You can always get clues. Be detectives, go out and knock on doors, and do all that you can to solve the mysteries because there's a lot of value there that you can see by actually being in the environment. You'll find a lot of good insights to leverage."

2 **Crash Pilots**: too often startups fall into the honey trap of being seduced by big logo customers with pilot agreements (discounted or even free) in exchange for the ability to validate the technology in the enterprise. These relationships usually start in the innovation departments, but your goal is securing a revenue-driving contract with the business unit. The problem is the internal politics between the internal divisions, as these types of pilots are often disconnected from the company's P&L.

Simpson encouraged, "I think it's really important to get an understanding of the organization you're selling to. You have your main contact, let's say in this case, an innovation person. You must be pushing and asking questions about where to go next. Who has to approve it? Find ways to get them engaged earlier, and if they won't get engaged, really understand if it is a budget constraint. Is it just bandwidth? Truly understand who your ultimate customer is, because it's probably not that innovative." Simpson has seen that too often cool tech becomes pet projects that never become purchase orders, this is a death spiral for a startup.

He advocated for startups to ask upfront, "For other examples of other technologies that are deployed at scale that have gone through this process?" Simpson has witnessed the most successful implementations of PoCs (proof of concepts) when the entrepreneur becomes known on-site, not just shipping the technology to the organization and waiting for an answer. "Find ways to get on-site to just get a feel and make yourself known. Most often when that happens, they're going to invite other people to the meeting, and you can start trying to expand [the PoC]. In other words, expand your understanding of the group that you're selling into," stated the CVC. Finally, he thinks founders must weigh the value of the relationship from the onset. "You're going to be investing a lot of time, money, and effort into figuring this all out. And if you know upfront it's only going to be a low six-figure type of recurring revenue, that's not really that meaningful. Then, you might be better off focusing your efforts elsewhere," Simpson commented.

3 **Champions Rule**: playmakers start by knowing all the positions on the field; in this case, check where your receivers are running as you head into the red zone; in business speak, build a close relationship with one risk-taker who is going to shepherd your solution across the corporation, even to their bosses. As Simpson coaches, "You need to find a really strong champion within the parent company because there will be plenty of opportunities to give up along the way. I was even told to give up along the way, so having conviction is the most important thing."

The former CVC at GM, Amazon, and Yamaha, Simpson has worked in a variety of cultures, but every organization will have a fixer. "There's always going to be folks that have a mindset of we have a problem to solve, and it doesn't matter if I solve it, or if I work with a startup to solve it. This is a critical pain point: we need to solve it, and I'm actively looking for solutions for it. That's the ideal kind of customer to align with and they exist at every company, no matter how big or how long they've been around," Simpson clarified.

"When I was at GM, I spent a large part of my first year looking at lidar technology. I looked at a bunch of different lidar technologies. I would talk to various people, and get various opinions, but the same name started coming up. Whether I was talking to a vice president, engineering manager, or another engineer, they all said, 'What does this person think?' Find that person that folks are listening to. They exist for every critical problem in an organization. And it's often not the highest ranking person. It's the key expert, the go-to person that you want to be working with," he cited.

4 **No Life Rafts:** it is important to remember that the CVC is not your friend or therapist; even though you are becoming intimately familiar with each other's lives, your contact is still an employee of your customer and investor. For example, I had a company that was strapped for cash and approached the CVC of a huge global conglomerate for more money. The CVC called me and asked if I thought the IP had value. They essentially gave a lifeline that enabled the corporation to acquire the company at a quarter of its value, or they would cancel their million-dollar contract and tank the startup. This is the dark side of the CVCs that every founder should keep in the back of their mind as they navigate these critical steps.

Accelerators + Incubators

Close to 3,000 accelerators and startup incubators throughout the United States offer startups a launchpad to propel their businesses forward with curated programming, capital, and mentor networks. While many of these platforms are geared towards pre-revenue, pre-product companies, today's cohorts include later-stage enterprises that need the resources and support to scale their businesses. The key for founders is securing a relationship that matches their go-to-market strategy and product objectives.

As a VC and mentor of three accelerator programs, I often refer startups to the application process of such partnerships for capital and resources, especially hardware businesses that need access to rapid prototyping and testing facilities. To understand the full extent of the opportunities and selection process for robot and drone startups, I spoke with Kara Jones, executive director of GENIUS NY. "Sometimes you see accelerators that are industry agnostic, but I think it's really unique that GENIUS is focused on one industry [robotics]. So that means that we have companies coming at different levels. This year, we had a team come in that had already raised $11 million. Yes, the initial money out of the accelerator is great, but they're coming into the accelerator to set up shop in Central New York, gain access to its [FAA] test site, and work with some of the teams that are already integrating into their system. So, I think accelerators, depending on which one, are not just for teams in the ideation phase," she promoted.

Jones further illustrated that the networking aspect of the co-working environment could be one of the most valuable fringe benefits, "Being surrounded by founders and experts in related fields, sharing insights and experiences—who knows what innovations can arise just from being part of that kind of community." Many of the early adopters of GENIUS NY's portfolio companies were internal. As Jones reminisced about Cohort 6's collaborative environment, "Josh Ogden from AVSS Drone Parachute Recovery Systems is a great example; he joined the program, connected with companies like Blueflite, and started collaborating with our fleet management partners. AVSS has been key in creating valuable connections within our teams."

As you research your accelerator options, you will see that each one will have its particulars. For example, GENIUS NY is backed by New York State to build up the economy of central New York, so they require that teams move to Syracuse, NY, for at least a year. There must be alignment between your mission and theirs. Jones unpacked her selection process, "Because our program is funded by New York State, our first priority is finding teams that are truly committed to relocating to Central New York and making it part of their growth story. Second, we need to see a strong solution paired with real market understanding. Too often, I come across teams with a product but no clear market or customer—they've built something intriguing but haven't fully thought through who it's for. That customer discovery process is invaluable, and it's something teams often overlook. Finally, it's about the people. We look for founders who are open to feedback, ready to immerse themselves in a new community, and willing to navigate the intricacies of the UAS market, from manufacturer relationships to regulatory engagement with bodies like the FAA. The teams in GENIUS NY that succeed here are those who take full advantage of the ecosystem."

This is further illustrated by the experience of Andrea Thomaz of Diligent Robotics. "So we had a relationship with Cedars Sinai in LA, and we went explicitly through their accelerator program," she recalled. "Honestly, when we joined their accelerator, I was thinking we were a four-year-old company, we were too far along for accelerators. We have a product that's ready for market, we need customers. But the CIO at Cedars convinced us that this is going to be what you need. He was absolutely right because, after that experience, we had the Cedar's stamp of approval, which was gold. In the six months after our Cedar's deployment, our pipeline quadrupled and all of a sudden we had this market validation and everybody was happy to talk to us."

While Diligent Robotics was not a GENIUS company, Thomaz's experience at Cedar's accelerator validates many facets that Jones shared about her cohorts in Syracuse. Thomaz explained, "One of the first things we did was get out in front of our customers as fast as possible. We raised a little bit of seed capital in 2018 and within nine months, we had a first prototype built. We were embedding ourselves in hospitals and testing both technical feasibility and product market fit. I think probably with hardware more than software, making sure that you get out in front of the customers as fast as humanly possible is critical. This was one of the reasons why we were able to get that product-market fit nailed down faster and not spinning our wheels."

While the value of VC is often money first and intangible support second, CVCs and incubators bring not just capital but design support. Thomaz's early deployments led to a fundamental pivot that enhanced the product's value proposition and sales. "We kept hearing about fetching and gathering that nurses were doing and so we really thought that's what they needed robots to do - go into a supply room, grab something with an arm, and bring it to a person. Because we were embedded with the teams, they also started asking, 'Could the robot transport this lab sample for me? Or could the robot go to the pharmacy and get this medication for me? Or the IV pumps that are stored all the way in the basement?' And that wasn't the core thing that we set out to do in our research trials, but it's the thing that people asked us to do. We kind of stepped back after that year of being embedded with hospitals and said, 'The things that we did in the hospitals, where did we provide the most value?' It's all of those things. And it turned out that was our most important focus," the healthcare roboticist affirmed.

Venture Debt

There are other forms of capital (many grants available to mechatronic innovators, which we will discuss in the next chapter). Still, venture debt and equipment financing are widely available. The irony is that to access these services, you need to already have cash in the bank and sales. Typical venture debt offerings from banks require at least $3 million in annual recurring revenue and $1 million in the bank. This is more of a post-Series A financing instrument to extend the runway. You also have to seriously consider if you want to collateralize the company or IP if you go into default. There is also equipment financing to enable hardware startups to secure inventory, which again requires sales and purchase orders to offset lending risks. These debt instruments often require steep interest rates that are encumbering to startups.

A good rule of thumb is never to allow your debt service to exceed 40% of your monthly income; I recommend keeping it under a quarter of your topline revenues. If sales stagnate pretty soon a company could be de-leveraged and go into default. When salesmen approach you about these 'attractive' accessible capital lending platforms, you should discuss it with your board and investors, as your shareholders are your ultimate stakeholders.

Founder Insights: F. Javier Diez, Amphibian Robots

Some ideas are so innovative that they are ahead of any market. One of the roles of the government is to fund research in the pursuit of invention and progress. Submersible drones that swim and fly are an example of a technology that has yet to achieve product-market fit but could be on the cusp of a novel new workflow for inspecting infrastructure and defense armaments. Read through the interview below with Professor F. Javier Diez of Rutgers University and begin listing an investment rationale for investors to seed his company.

Background

Following the news of the implosion of the Titan submersible in 2023, I reached out to Professor Diez for his comment on the rescue mission and the role of robots. The aerospace academic is also an entrepreneur of a novel drone technology company that can fly and swim autonomously within the same mission. As he explained, his approach could have saved time and money in ascertaining the same unfortunate answer, "I think a variation of our Naviator drone could go down to 12,000 in the near future. No problem. So now imagine sending a 20-pound [robot] down to 12,000 feet. You can do this in a couple of hours. You just throw it overboard, or you fly, you know you don't need to bring in a crane, a gigantic ship, and all this very expensive equipment just to do that first look." Diez continued, "We could have been there in a couple of hours. So of course, you know there's more to it. But I was just saying that in the long term, I can see how very small robots like ours for search and rescue could be huge. We are doing some work. We actually put some proposals with the submarine community. I think this has a huge application because again, these 20-pound [drones] are something you can deploy from anywhere, anytime."

Invention

In breaking down his invention, the drone CEO elaborated on the epiphany in his lab years earlier by overcoming the conventional wisdom that an uncrewed system operated in two modalities (marine and air) required two separate propulsion systems. He further noted that two propulsion systems were very inefficient regarding burning energy and functionality.

"And this was, I would say, a mental barrier for a lot of people, and it still is when they see what we put into it," he explained. The professor continued to elaborate on how he first had to overcome so many industry naysayers, "I brought this to some folks at NASA, and everyone was saying, it's not going to work. And then when you look at what's behind the propeller design and the motor design, you realize that we cannot be living on an edge. We designed propellers for a very specific condition, which is air." However, the innovator challenged the status quo of the aerospace community with the question, "Can you design propellers and motors for water? And it turns out that you can." He deconstructed his lab's research, "So if you look at the performance curve for air, and you look at the performance curve for water, they intersect, and if you do it the right way, you can be efficient in both places. So that was the breakthrough for me to be able to show. And we actually show that you can design propellers that can be efficient in both air and underwater." After sharing insights into the design, he conveyed to me that the programming of the flight controls was the next hurdle to overcome.

"The next challenge is the transition. So we worked very hard from the very beginning on that transition from water. We actually have a patent on this, and it's really the heart of our technology. I call it dual-plane propulsion. You have propellers on the top plane and propellers on the bottom plane. So when you're on the surface, the bottom ones are in the water, and the top ones are in the air. So the bottom ones are like when you have a baby, and you are pull-swimming. Babies are not very good at swimming, but if you put your hand on their bellies, all of a sudden, they become great swimmers. So think of it as the bottom propellers. When the vehicle is on the surface, the bottom propellers keep it very, very stable. So now that you have that stability, the top [propellers] can work together to get [the drone] out of the water. So that's one key part of how we accomplish the continuous transition. You can go in and out 100 times," bragged the Professor.

Potential Markets

Diez's company, SubUAS, is not a theoretical concept but an actual product currently being tested by the US military and is looking to expand into commercial markets. "So we'd been a hundred percent with the Department of Defense. They really supported the development of our Naviator technology," stated the entrepreneur. He is now itching to expand from a Navy Research-funded project to new deployments in the municipal and energy sectors.

"We have done a lot of different types of inspections related to ship pylons. Now, we have the Department of Transportation interested in this technology," said the startup founder. "What I realized over the last year or so is that defense has its own speed. You cannot really push it. There is a specific group now in defense that is encouraging us, but it takes a couple of years," he quipped. Optimistically, he envisions being profitable very soon by opening up the platform for commercial applications. "Now we're starting to see the fruits of that [effort]. I can tell you that we got approved in Europe to do an offshore wind turbine inspection later this summer," he expanded. However, he is most excited about bridge inspections, "We have over half a million bridges in the USA. And, like at least 50,000 to 200,000 have something seriously wrong with them. I mean, we're not doing enough inspections. So having a vehicle like the Naviator that can look at the underwater part of the bridge is huge."

Several companies have also approached him in the energy industry. "And then there are a lot of interesting assets within oil and gas, but we are discovering this. It's kind of almost like a discovery phase because nobody has ever had the capability of doing air and marine," he elucidated. He described that there are many robots like ROVs (remotely operated vehicles) inspecting rigs on the marine's surface and aerial drones looking from the air, but no one is focused on the splash zone [where the two meet] as they never had dual modality before. He further illustrated the value proposition of this specific use case, "Nobody gets close to the surface. So they're saying that that's a huge application for us." Diez imagines replacing tethered ROVs altogether as his system is easier (and cheaper) to deploy.

Today, SubUAS's business model is on an inspection basis. Still, over time it will center around data collection as they are the only waterproof aerial drone on the market that can swim. "We go to the bridge inspectors, and we work with them to simplify their lives, and at the end of the day reduce the risk for the diver. So they know what we are doing is making their lives easier," said Diez. However, that is only the tip of the iceberg, as he expressed, "It's not so much about the hardware or the sensors, but the data you collect. We think cloud services are huge, as it could allow you to sort and analyze it anywhere." He concluded by sharing that his next model will utilize a lot of artificial intelligence to interpret the condition and autonomously plan the missions accordingly.

📖 *Assignment*: New disruptive products have so much potential that startups often die because of a lack of focus. As a newly hired executive for SubUAS to dun its strategy and business development, break down your plan for creating a dual-use company with two divisions, military contracts, and commercial sales. What would be the low-hanging fruit to prove the technology and expand within an industry? Also, as this is a pre-revenue, prototype-ready company, think through government non-dilutive grant opportunities for Diez and his team to use as a funding instrument ahead of its first large order.

7

GOVERNMENT AS AN INVESTOR AND CUSTOMER

Flora and Fauna

The government is also an accelerant of capital, development, and potentially big business. Defensive innovation accounts for close to a billion dollars of investment from American taxpayers into funding research that is too early for private capital. As Nic Radford of humanoid startup Persona AI, boasted of his work leading NASA's humanoid Valkyrie project, "The government is a great place to develop IP. They're very favorable IP rights for companies." Besides the space agency, every government body has a funding arm navigating this labyrinth of bureaucracy, which requires diligence, tenacity, and lots of friendship with civil servants.

The Endless Frontier

I first met Retired Colonel Patrick Mahaney at New York University's Endless Frontier Labs, practically named after Vannevar Bush's paper post World War II.[1] The world war became a catalyst for innovating through a close partnership between the commercial sector, government, and academia. Charged with finding a way of harnessing the collaborative environment between three very different branches of invention, Vannevar Bush was appointed by President Roosevelt and then Truman. He drafted a paper called "The Endless Frontier." His words are valid today: "Basic research is the pacemaker of technological progress." In the 40-page report, Bush laid out a plan that would create the eventual government grants system across every agency, which would later be bolstered by the Small Business Administration and its Small Business Innovation Research (SBIR) grants that are utilized by the Department of Agriculture, Department of

DOI: 10.1201/9781003508410-8

FIGURE 7.1 Nic Radford and NASA's Valkyrie humanoid robot.

Photo Credit: Evan Ackerman/IEEE Spectrum.

Commerce (NOAA), Environmental Protection Agency, Department of Education, Department of Energy, Department of Transportation, and the Department of Defense (DoD). In addition, the SBA also administers the Small Business Technology Transfer (STTR) grants, commonly known as America's Seed Fund. The difference between the two non-dilatative funding mechanisms is how the money is split between partners and academia, essentially an SBIR gives more flexibility, while an STTR requires that 40% goes to a research institution (Figure 7.1).

Col. Mahaney on Working with the Military and Accessing Battlefield Feedback

The views expressed below are exclusively Retired Colonel Patrick Mahaney's sole opinions and do not necessarily reflect the position of the U.S. government or the US Department of Defense.

Col Mahaney today advises the DoD, including the Defense Innovation Unit (DIU), applying his Green Beret battlefield experience, to focus startups on countering what he called the "democratization of destruction." As the former soldier related, "You've got with the rise of computational power, portable electronic devices and put it all together essentially every squad, every small unit infantry force suddenly has their own Air Force. These capabilities are now also in the hands of non-state actors and are continuing to evolve. That's extremely significant. Likewise, I recommend l looking into the challenge areas that have specialized needs, such as in urban areas, and certainly subterranean environments. Given the rise of technology, now you have the ability to do some remarkable things that were unthinkable before."

It was around five years ago when robotics and sensor technology really began to take off. Congressional defense budgets planned for a much larger DoD spending, and it eventually filtered down to public safety municipal allocations for police and fire departments. "That's the need that drives us all, and what we are seeing from these other agencies, be they municipal or federal. And what are our allies and partners are seeing, and especially with what's going on the battlefields around the world, which since 2022 has skyrocketed," detailed Mahaney.

As the war veteran of Afghanistan and other conflicts advised his army of founders: "So the first thing you need to recognize is that you're dealing with these behemoths of government. Nothing moves fast, and very, very few things operate fast in Washington. The point is you have to understand that with the military and others in the government, it's going to be a longer-term prospect. You have to have some patience, and you have to understand that you're dealing with bureaucracies. I would say you need to find a champion. The champion certainly could be an office within a department or a unit within an organization. At the end of the day, bureaucracies are made for people, but within those bureaucracies, you've got some people who have a true sense of urgency, and in many cases have the authorities and the funding to make things happen. You want to be able to figure out who that person is going to be, or at the very least what the office is that is going to be able to help you navigate the process."

He maps out a plan for founders, to start relatively small, by illustrating the value in testing out their technologies with research and development departments. As he shared, "You normally have to prove what you've got in dealing with us. What we're really good at is iterating. If you get in with the right people, you can iterate. You get user input for free. The government's not going to charge anything for this." This is a big difference between private capital, which costs founder's equity, and public funds that come with design contracts with user feedback. As Col Mahaney reiterated, "This is the difference between us [the DoD] and the VCs. I describe it as a blue-sky approach. We tend to take an approach that is not laser-focused on extracting maximum value from technology, in this case, robotics. We have money and authorities that are meant to solve certain problems. So with us, even though you want to show your differentiation. You also want to be able to show that your technology could be used in applications. We see that a lot in the emergency medical field. We see that a lot in situational awareness, in anything that can help provide a common operating picture. That it is better, faster, clearer, and that can work through the data and get into spaces that we couldn't see before."

More than a Purse

The retired colonel thinks a balance of venture capital and government funds could offer the best approach. He recommended, "You want to be able to balance the VC approach to maximize the value and couple that with the broader view and due diligence of the government, and be able to say to investors, 'Look, the government is working with me, and my company is a startup.' Similarly, there is the importance of non-dilutive funding. It's intuitive, and yet completely unknown to many startups. It is surprising to me that many startups do not realize that they can get government grants. In addition, with SBIR and the STTRs, it's not just the dollars that you get, but there is value in going through the process itself, painful as it may be. Once you're actually working with an end user in the DoD or a program office that is working on the topic, then you start to get some pretty good feedback." He warned, however, that some founders "Take that too far, and become essentially part of a SBIR mill. They're just looking for free money from the government to do science projects, which is interesting, but it won't turn into anything significant."

Dr. Henrik Christensen: A Roadmap for US Robotics

Dr. Henrik Christensen drafted "A Roadmap for US Robotics" every four years.[2] He presented the most recent fourth edition to Congress on April 30, 2024. The report stated, "Robotics technology will transform society and is likely to become as ubiquitous within the next decade as computing technology is today." Most importantly, this document presents a guide for ripe areas that are in need of mechanized disruption across the United States. While the complete text of the Roadmap is easily accessible for free online, below are some worthwhile highlights for founders looking to ideate new robotic solutions the government may have the mandate to fund:

Reshoring and Onshoring:

- The ecosystem around collaborative robots (cobots), such tasks as bin picking, pick-and-place, and quality assurance as the report stated,

"Robotics presents great promises in redefining future manufacturing, alleviating the labor shortage, and improving the resilience of the U.S. supply chain."

- The Roadmap highlights a few areas: "There is a need to improve the adaptability and reconfigurability of manufacturing robots. For example, soft and compliant graspers are needed to handle fragile and irregular materials such as in food processing, and specialized end effectors are required to operate in landfills and process bio-waste in hospitals."
- Finally, it stresses how technology can advance the U.S.'s infrastructure: "Robots can also play a key role in manufacturing in extreme environments where manual manufacturing is challenging, such as space, underwater, or under extreme weather conditions."

The "Now" Economy, as the report stated, "Robotics can serve to increase the versatility of available resources, whether that is the customization (or fabrication) of goods or the performance of services." Opportunities in this space include:

- Shipping and logistics, especially autonomous vehicles around last-mile delivery.
- Telepresence solutions that can offer remote services.
- A gig automated workforce that works on on-demand automated to fulfill the void left by human labor shortages (i.e., harvesting of agriculture).

Feeding the World: Christensen's document also investigated the following: "Over the last couple of decades, there has been a growing interest in using robots as part of food processing. The spectrum covers all aspects of food processing, from planting seeds in the ground to weed removal and picking mature fruit/vegetables. Robots have generally been applied to precision agriculture, weed control, nursery automation, and harvesting. Precision agriculture is used to monitor crops, collect data, and apply fertilizer."

Regarding *Aging Society* it divulged opportunities for automation "As a transformative force in this landscape, offering a significant opportunity to reduce costs while making treatment and support, smarter, less invasive, and more personalized."

In describing new advances in *Surgery*, the report published, "Robotic technologies enabled by increased funding are needed to simultaneously improve clinical outcomes, making treatments smarter, less invasive, and more personalized, and to contain and reduce healthcare costs by facilitating more effective and earlier treatments and fostering increased competition in the healthcare sector."

More specifically Christensen highlights the challenges of **Aging in Place** for society, with robotics providing:

- Physical support, he stated: "These physically assistive devices should be part of a larger aging-in-place support/tele-support system, including digital health monitoring and pre-clinical diagnoses. This integration is essential for facilitating the safe, cost-effective, and pandemic-resilient delivery of rehabilitation, nursing, and care services directly to individuals' homes. The same systems that provide physical assistance can also enable cognitive benefits."
- Cognitive support, the roadmap delineated, "Robots can engage in many daily interactions with the elderly user, providing social and cognitive stimulation, reminders, and information, while also analyzing the user's speech, affect, pose, walk, activity, medication adherence, etc., to look out for early signs of cognitive, emotional, and/or mental health decline."

Regarding **Housing and Infrastructure**, he stated, "We see an opportunity for the use of robots to increase productivity in construction while reducing worker injuries." The congressional report breaks down the opportunities in this sector:

- Heavy labor, as described, "Some companies are already exploring the use of robots to help with bricklaying, moving heavy items on construction sites, and framing houses."
- Inspection and repair recorded how "Robots could conduct inspections and to autonomously repair, pave, grade, stripe and mark roads, all under the supervision of human workers who would move from physically intensive jobs to more supervisory roles. Robots can also inspect and repair other types of transportation infrastructure like subways, trains, airports, and ports." *Note: CivRobotics has a product called CivDash for road striping.*
- Waste management, as it explained, "Robots could sort waste from recycling and organic matter to reduce the amount that is burned or goes into landfills, with this sorting both at local pickup locations and larger sorting centers."

The Roadmap aims to tackle challenging issues like climate change, as it illustrates how robots are furthering **Renewable Energy**, as it suggested: "In the area of natural resources, robots can help with renewable energy (e.g., construction and maintenance of wind and solar farms), land and mineral management (e.g., mining), and water (e.g., repairing and replacing aging pipes, desalination)." The following is a list of immediate ways for automation to advance the

- On sustainability, it disclosed, "Opportunities for robotics to support sustainability span multiple areas, that range from manufacturing applications to agriculture, and include development of renewable energy, maintenance of aging infrastructure, reducing food waste, adapting to a changing climate, as well as quality of life for workers."
- Regarding manufacturing, the report outlined, "Sustainability of manufacturing requires the ability to automate the handling of hazardous materials to avoid potential harms to workers and the environment (e.g., the lithium required for batteries to electrify transportation)."
- In advancing alternative fuels, the roadmap set forth, "The development of solar farms ranges from the design of the layout of the farm to driving the piles and constructing the solar cells [similar to CivRobotics]. To maintain the solar farms, cleaning, mowing, and management are needed and can be provided by robots. Similarly, robots can play a major role in the prospecting, construction, and maintenance of wind farms, geothermal, wave every, nuclear and other."
- Christensen outlined how technology can address climate impact, "Robots can help in mapping fires and directly intervening to extinguish or contain them. The increase in the strength and frequency of hurricanes and tornadoes also requires support from robots in mapping the areas affected, and helping in search-and-rescue operations. Automation is also critical to help farming adapt to a changing climate."
- In alleviating the current maintenance issue, it stated: "Robots can be used to inspect and maintain aging infrastructure, like roads, bridges, buildings, dams, railways, and power lines."
- Recycling programs should continue advancing, said the report: "Robots are required to recycle hazardous materials that are key for environmental sustainability (e.g., lithium batteries). Furthermore, new automation technologies will be key to moving the recycling or reuse of materials beyond those that can be easily automated (e.g., separating metals and paper products) to increasingly complex products (e.g., electronic devices)."
- The report presented quality of life opportunities, "Robots can also improve the safety of workers in manufacturing environments, agriculture, and other roles such as fighting wildfires, all of which are key for improving the sustainability of the economy by improving the quality of life of workers."
- The next wave of transportation is further advanced by the "Use of autonomous or semi-autonomous vehicles can optimize transportation systems... There is a lot of potential in the utilization of intelligent vehicles from micro-mobility to trucks to reduce the environmental impact."

Expanding Frontiers: "The realm of space offers unprecedented opportunities to expand our technological and economic horizons." The list below presents more opportunities for founders looking to advance frontier technologies:

- Space mining, the roadmap offered, "This involves the extraction of valuable resources from celestial bodies such as asteroids and the Moon. Key elements targeted include rare metals like platinum, iridium, osmium, and palladium, which have critical applications in various industries."
- On-orbit servicing and repair it stated, "This sector focuses on maintaining and enhancing satellite operations through refueling, repairing, and upgrading while they are in orbit. Such services can significantly extend the operational lifespans of satellites and enhance their functionality, offering a sustainable solution to manage the growing satellite populations in Earth's orbit."
- Christensen detailed new areas of space manufacturing and recycling, "As space activities increase, so does space debris. Recycling this debris to recover raw materials presents a unique solution that could support manufacturing directly in space."
- Space exploration and infrastructure are advanced by "Developing technologies for habitation and work on the Moon is a critical aspect of modern space exploration."
- In-Situ Resource Utilization (ISRU) could offer roboticists opportunities, as it covered "ISRU focuses on using local resources in space environments to support human life and enable the construction of space infrastructure."

In discussing the importance of the roadmap in driving real-life applications, Christensen explained, "For the roadmap, we sort of went in and said here are the business cases, how do we make money at the end of the day? We don't do research, just to do research. We research to really understand what the business cases are based on and where the obstacles are that are not allowing us to make progress on that today." This thought process is baked into how the government utilizes its vast funding arm to support startups that could be too early for private capital. Dr. Christensen always asked, "Where are the interesting problems, and how can I address them?"

Today, the roadmap serves as a blueprint for the USA to invest in technologies and evolve, even though some policymakers seem to ignore its current suggestions. To illustrate the power of this document, it has been successful in critical points in America over the last decade and a half:

- 2008 – during the Great Recession, policymakers look to retool American manufacturing to better compete against Far East manufacturing;
- 2011 – utilizing mechatronics and lessons learned from the Fukushima meltdown for modernizing disaster, search and rescue, and recovery programs;
- 2016 – with the trade war with China, investing in collaborative robots and autonomous vehicles to accelerate reshoring and near-shoring priorities;
- 2020 – using robots en masse to address supply chain shortages, especially with personal protection equipment, brought on by the onset of COVID and
- 2024 (today) – highlighting the opportunities at the cross-section of deploying AI-enabled machines, "How do robotics and AI come together to build some sort of conference."

Accessing Resources via Robot Clusters

The world has a growing presence of robot clusters or community centers for roboticists that partner with local governments and industry to help innovators raise capital, generate sales, and utilize municipal resources. The Clusters are a relatively young endeavor, and every day, I hear of new organizations that act unofficially as a cluster. The growing list of official cluster members can be found on The Global Robotics Cluster (GRC) website, today's presence across the globe includes[3]:

- Daegyeong Robot Enterprise Promotion Association (REPA - Korea)
- MassRobotics (USA)
- USA: Silicon Valley Robotics (SVR - USA)
- COBOTEAM (EU)
- France: FFC Robotique (FFCR - France)
- Russian Association of Robotics (RAR)
- Singapore Industrial Automation Association (SIAA)
- Malaysia Robotics & Automation Society (MYRAS)
- Israeli Robotics Association (IROB)
- Turkey Robot Association (ROBODER)
- Spain Robot Federation (Hisparob)
- Mechatronics & Robotics Society of the Philippines (MRSP)
- The National Office for Innovation and Technology Transfer (NOITT)
- New Zealand Robotics, Automation, and Sensing (NZRAS)
- Robotics Society of Romania (RSR)
- The Robotics Society in Finland (RSF)
- Asia Pacific Assistive Robotics Association (APARA)
- Pakistan Robotics Cluster (PRC)

- KazRobotics (Kazakhstan)
- Foundation for Education Initiatives Support (FEIS - Kyrgyzstan)
- Australian Robotics and Automation Association (ARAA)
- All India Council for Robotics & Automation (AICRA)
- Aquitaine Robotics (France)
- Taiwan Automation Intelligence and Robotic Association (TAIROA)
- Lithuanian Robotics Association (LRA)

In addition, other unaffiliated clusters are operating outside the GRC, including:

- Pittsburgh Robotics Network
- New York Robotics Network
- Canada's Mitacs
- Israeli Intelligent Robotics Center (IIRC)
- Denmark (Odense Robotics)
- Uzbekistan Robotics Federation
- Pakistan Robotics Cluster (PRC)
- Taiwan Automation Intelligence and Robotics Association (TAIROA)
- Robotics Association of Nepal (RAN)
- Armenian Robotics Association (ARA)
- Cameroon - Sparte Robotics Cluster

As the executive director of Mass Robotics, Tom Ryden explained, "We're looking to find people who are thinking of new ways robots can solve different challenges that haven't really been addressed before. We look to make sure that it's an area that we think is of interest and that we can support. We want to make sure that we're providing value to the startup. This includes helping with fundraising to customer development/commercialization strategies. Founders must look at the cohorts before applying to avoid any conflicts."

As Ryden described, "We also don't take direct competitors, which when we were small was easy; now that we're 80 plus companies, it's a little bit harder. We don't because we're an open, shared community. We don't take companies who are directly going head to head and battling over the same customer."

He shared with me how construction robotics is now very important for Boston as they have some of the largest contractors in the world based in the New England hub. "We have a few construction companies here in Boston that are really engaged in looking at technology. They were meeting with one of our startups and we had a chance to say, 'Well, have you seen this other startup? Have you talked to this startup?' And so for them, it was great. They were in to see one startup, but they could actually meet three or four," Ryden excitedly shared. As he summed up, "We look to make sure that start-ups are going to be a good fit with our community. I think that's the number one thing that we look for."

Founder Insights: John Ha, Robot Waiters

Today, the labor shortage and operating overhead in running casual dining establishments have pushed restaurant owners to adopt new automation technologies feverishly across the globe. While many entrepreneurs have opened and closed robo-themed establishments, Bear Robotics has successfully introduced fleets of autonomous servers to ferry food and dirty dishes back and forth from kitchen to table. Read through this early interview with its founder, John Ha, from 2017 of the company that has since raised over $185 million and delivered more than 300 million dishes to diners since its inception. Be sure to note the opportunities for the company to scale.

Background

According to post-pandemic trends, restaurants nationwide are experiencing high employee costs associated with increasing waitstaff churns and labor shortages.[4] Server discontent translates to bad customer service and the loss of substantial revenue. This shift significantly affects smaller and family-owned dining establishments. At the same time, unions are pushing larger chains within the hospitality sectors to increase pay and benefits, leading to a faster push towards automation technologies in both the back of the house (the kitchen/stockroom) and front of the house (dining room/cashiers). One company that is reaping the benefits of these macroeconomic trends is Bear Robotics, an autonomous food runner platform.

My 2018 interview with John Ha, founder and CEO is still as relevant today as it was then for a peek into his ideation and market disruption that now works with Fortune 500 Brands and LG Electronics as its largest shareholder:

Ha has been making headlines with his game-changing product: "Penny – The Runner Robot" [note: in 2020, Penny's name was changed to Servi]. Ha shared with me the evolution of his whimsically named mechanical waitress. He started his career at Google, working long hours. In between coding, the former engineer would eat dinners at a small Korean restaurant minutes from Mountain View. Ha eventually bought the Kang Nam Tofu House, dreaming of starting his own chain of casual Korean dining experiences.

"I experienced the challenges faced by many operators in this industry. When my employees would show up late to work — or not show up at all — I had to step in and carry the load. And yes, that meant cooking, dishwashing, serving, bussing, hiring, etc," as Ha sighed. Then, after two years of slaving away, "A light bulb finally flashed over me when I was knee-deep into this. I said to myself: there has to be a better way to run these establishments," described Ha. After testing several concepts, he finally landed on the idea of a "food runner robot." In his experience, he continued, "This is a simple task that restaurants can use to take a burden off of servers." He continues, "From a front-of-house perspective, running food from one location to another does not add tremendous value. So, why not automate this?"

I asked Ha if he thinks Penny could mean the death of the food service profession, and he retorted, "This was less about replacing servers with a robot and more about changing the nature of a server's daily work." Using the Tofu House as his laboratory, Ha found that his servers spent less time running marathons of shuttling food than with customers. Proving his thesis, Ha boasted that his "servers generated an 18% increase in tips," after Penny took flight. He also thinks that another side benefit of robots will be longer employee retention by making their work more enjoyable. As Ha elaborated, "Imagine yourself walking or running five to seven miles a day juggling multiple trays of food and drinks in a narrow and crowded environment. If you can picture this, then also think of how physically taxing it is."

In discussing the competitive landscape of similar offerings with Bear Robotics' Chief Operating Officer, Juan Higueros, says he is not discouraged as Penny has already successfully served over 25,000 satisfied customers [and as of 2024, over 2 billion steps saved]. Higueros stated that the real opportunity for his company is to partner with restaurant brands, such as Darden's Olive Garden and Yum's Pizza Hut, to deploy its "Bear Operating System for remote fleet management and on-site operation," on a global scale. In the meantime, Ha and Higueros' team is busy perfecting Penny's AI-human interface to amplify the value proposition.

Assignment: In reading the original 2018 interview with John Ha, now go to the Bear Robotics website and see if he has stayed true to his original mission. Be sure to notice what adjacent markets Bear has entered into, and the types of clients it services (small or large). Now, you are charged with maximizing the Customer Lifetime Value (CLV) of those premium accounts. What steps would you undertake to broaden the role of a Customer Success team in achieving this goal?

Notes

1 Bush, Vannevar, and United States. 1945. Science, the Endless Frontier: A Report to the President. Washington: United States Government Printing Office.
2 Christensen, Henrik . 2021. Roadmap for Us Robotics - from Internet to Robotics 2020 Edition. S.L.: Now Publishers Inc.
3 CLUSTER, ROBOT. 2022. "글로벌로봇클러스터, 글로벌로봇클러스터홈페이지, GLOBAL ROBOT CLUSTER." Higrc.org. 2022. http://www.higrc.org/main/.
4 Fuller, Joseph, and Manjari Raman. 2023. "The High Cost of Neglecting Low-Wage Workers." Harvard Business Review. May 1, 2023. https://hbr.org/2023/05/the-high-cost-of-neglecting-low-wage-workers.

8

SCALING UP GROWTH MOUNTAIN

Summiting

As the great twentieth-century philosopher Theodor Seuss Geisel declared, "Congratulations, today is your day; you're off to great places, you're off and away."[1] This statement echoes a founder's excitement after closing their first funding round. However, Dr. Seuss tempered his earlier statement: "I'm sorry to say so, but sadly, it's true that bang-ups and hang-ups can happen to you."[2] For a startup, the scrappy pre-financing attitude must continue through its early development, scaling up smartly and not spending too liberally. Sometimes, raising the first round is easier based on the dream versus post-market when a company has sales and financials to analyze. Missing your revenue plan could harm a company's growth, so a good founder ensures they raise at least enough to last 18 to 24 months to work out the kinks.

Below are some helpful guiding principles tips from successful robotic founders who have scaled their companies through to exit:

Rule 1: Be Frugal

Mick Mountz, founder of Kiva, commented on the success of Kiva's trajectory, "We got lucky at Kiva that our business model supported very little need for outside capital. We got to profitability by raising only $33 million in venture capital, and we were cash flow positive for all of 2010 and all of 2011, two years prior to Amazon, even buying the business. The reason we were successful is because when we would sell a $5 million system, it carried high gross margins, that would throw off cash that helped us grow the business without having to take so much outside capital. Essentially, booking

DOI: 10.1201/9781003508410-9

new business and bringing cash in the front doors. We had to be very careful with our dollars between the Series B and the C, and the C and the D."

He boasted about his frugality, even cheaper than Jeff Bezos. "When we got bought by Amazon, one of their core values at Amazon was frugality, and I thought that they were spending lavishly compared to the way we spent money." Cash management is a CEO's #1 job, but I've seen too often entrepreneurs surprised when they have weeks of cash left. Just because you have money doesn't mean you need to spend it, as a rule, I tell founders, ask yourself if the investment of capital will lead to more sales, if it doesn't, then there is little rationale to risk it. It's that binary of a decision process.

Rule 2: Align Your Team

Mountz's co-founder Raffaello D'Andrea adds that in addition to cash management, the job of a scaling company's CEO is keeping team alignment while staffing up. "My job as a CEO is really to have a vision for the company. Make sure that we have alignment. I think that is my biggest role. Right alignment is not just for our employees, but with clients and investors," D'Andrea professed.

He illustrated how his role has changed from bootstrapping product development to the C-suite of managing people, "Unfortunately, I don't get to write code anymore, but we have a lot of really good people that can do it better than me." This view was shared by Theobald of Vecna, who said, "You've really got to realize that building the right team is probably one of the most important things. A lot of our success was built around just finding an incredibly talented and committed team of engineers that you know we're uniquely capable and making things happen that other people weren't, you know, and the other one, of course, deals with money and financing." A good founder builds a hive that attracts quality talent like honey to the mission; at the same time, it is the CEO's job to delegate and delineate responsibilities across the organization to keep everyone on task toward this goal.

It's critical that, as CEO, you also respect the chain of command. For example, many great executives situate themselves in the middle of the action versus in a quiet corner office. This access could imply that you are available to answer any question; however, be on the lookout for junior people going around their supervisors and walking directly to your desk with ideas. After listening to their suggestions, don't reply (even if it's fantastic), but instead encourage the employee to go to their supervisor with such comments. Empowering your managers to oversee their team gives the organization clearly defined roles and autonomy to achieve its objectives. If you assert yourself in the hierarchy, everything breaks down, and all the authority below you becomes diluted.

Starting a company is all hands on deck, with everyone doing whatever it takes to drive early sales. Running a business post-inception requires coordinating teams and exchanging ideas. The challenge for many companies is scaling up as the team gets bigger so does the bureaucracy. Jeff Bezos of Amazon had the two-pizza box rule to avoid bloated departments and endless meetings. Bezos means that teams should not be any bigger than the number of people who share two pies.

Ryan Gariepy of Clearpath Robotics shared how his startup organized itself initially to set the foundation for the company to scale between his co-founders, Matt Rendall, Pat Martinson, and Bryan Webb: "We were comfortable with rebalancing a lot of our responsibilities as we scaled. There was a specific point I remember when we had challenges with our product strategy, and we had challenges with our sales numbers at the same time, and at the time both product and sales reported into Matt [Rendall]. We sat down for dinner and said, 'We really need to fix all of these things, and Matt doesn't have the mind space to fix both product and sales. He's very clearly better at sales and he's just a natural salesperson.' So we said, 'Okay, he'll take on sales leadership, and then Ryan will take on product.'"

He illustrated how this early delineation of responsibilities led to a more streamlined leadership team, "We realized it really improved decision-making. I think that's important for a robotics company that we ended up having one person where the buck stops at getting a product that works for customers out the door." As he explained, "We were able to make very quick decisions even in the face of uncertainty. 'For example, if it was very expensive to make a robot that runs at 2 meters per second, but very cheap to make a robot that runs at 1.5 meters per second, and both were reasonably acceptable by our users, we could converge on those combined product/technical decisions very fast.'"

While scaling up, it's important that an organization doesn't lose sight of the user feedback loop, as customer success is critical to growth. Clearpath implemented internal processes to evolve products and sales initiatives that were born from field experience. As he explained, "When customers are asking for things, make sure that your team is hearing that, and not guessing what customers want." The leadership team would communicate regularly, not just when things went well but when things went awry. He recalled, "Matt was a very strong customer advocate. Anytime we were on the verge of losing a deal or we were providing a poor customer experience, we definitely all heard about it. Part of it was having those communication loops in place. We would meet as an executive team once a week and just talk about customer sentiments and site metrics." Great companies put customers at the center of the equation, ahead of technology and investors.

Rule 3: Set Priorities

It's not enough to focus on the executive team, it has to filter down to every department. As Gariepy outlined, "You can't always scale by just having people talk to each other, there should be rules of thumb and best practices. A key practice we did for product management was putting in place certain guidance on the value of a deal versus the amount of roadmap disruption it could 'buy' to reduce the amount of urgent engineering team meetings. If you're coming in with a total contract value of 'X,' then this is roughly the number of roadmap changes you can get for that." He further explained that it doesn't mean that they wouldn't talk to customers about minor pain points at all, as feedback drives innovation, "At the same time, we had different parts of the organization that's always trying to understand what the customers want by talking directly to them. As our product team grew and we established a better rhythm, we eventually set in place a rule that the moment a customer would ask for features a product manager would be called to avoid any sort of broken telephone. […] It was more important for us to bother the product managers a little bit more to make sure that they had real-time information of what the customers wanted."

Rule 4: Plan Resources

As Rodney Brooks, serial founder of iRobot and Rethink Robotics, elaborated on his current priorities for his latest startup, Robust.AI, "We have to figure out ways to grow faster and faster and faster. We can have direct relationships with just a handful of you-know customers, and still we can't meet the demand. Once we get that under control, then we'll probably start looking for broader distributed distribution mechanisms. Right now, we are production limited."

Brooks continued to outline his current growing pains, "We're not trying to build out a large distribution network right now, because we can't keep up with orders. We don't have a product because we sell everything we can manufacture to current customers." Brooks cautioned that it is critical for growth that the startup is not too reliant on one customer but has a healthy distribution across the industry, as over-dependency can lead to failure. "You don't want to be a mono customer. A whole bunch of robot companies got burnt by Walmart about 3 years ago, At different stages of the deployment. Some only had two robots, some had 50 robots, some had 300 robots. And then one day, Walmart said, 'We're not doing this anymore.' And they [the startups] were all totally toast. So, we do not do mono customers," warned Brooks.

Rule 5: Follow Leadership Principles

Amazon is one of the largest companies in the world but still has a startup mindset regarding fostering a culture of innovation. The leadership principles

Jeff Bezos set for his organization in his first shareholder letter in 1997 stated its core belief to "Obsess Over Customers," which led today to 16 guiding ideas for its 1.6 million plus employees. Paolo Pirjanian, founder of Embodied Robotics and former CTO of iRobot, affirmed, "It's phenomenal that he [Jeff Bezos] has done that. It's easy to maintain that in a small team, but in a large team, I don't know. To be honest, we didn't have as large a team as theirs. I don't know how they [Amazon] do it, but they do it really well."

When Pirjanian sold his company Evolution Robotics to iRobot and became its CTO, he managed to preserve its innovation culture as it grew at hypersonic speeds. "It's a big job, big title. On the other hand, I, personally, also don't like dealing with politics and stuff like that. I like to be in a very honest open environment where people can speak up their minds and not be afraid of things backfiring or being misunderstood, and so on," he stated. As he elaborated on the initiatives he set out for the growing public company, "I drove the company really fast forward in terms of the pace of innovation. And we were moving really fast and breaking a lot of things. It required restructuring. It required spending a lot of time building a new culture, which was a culture of innovation. There was the fear of making mistakes with a lot of processes in place. People are afraid of making decisions because they're afraid that it would be the wrong decision. They make a mistake that slows down everything. We tried to remove a lot of those barriers because the culture and mindset had to change." Pirjanian principles became, 'make mistakes and learn as a way to innovate.' As he said, "And that changed everything, we went from product development cycles that would take 5 to 7 years to less than two years."

Nic Radford, who built the undersea robotics startup Nauticus into a 120-person organization that went public in 2022, compared his journey in scaling up his company to breaking the sound barrier. "The velocity of companies is no different. As they progress, there will be times of instability and shaking, but if properly designed and managed, they are able to punch through to the other side, finding smoother air. However, as the shock waves develop on the team's aerosurfaces, you'll quickly find if you're flying a company with swept wings," Radford declared. He recalled how he managed his own setbacks in raising his critical growth equity round, "During raising our C round for Nauticus, I pitched 278 VC/CVC/PE/Family Offices. All nos. But the 279th wasn't. Nor was the 280th. The fortitude and positive mental attitude you must maintain are critical and cannot be overstated. It's way too easy to get cynical, discouraged, and want to say, 'Fuck it all.' You can't. Further, you need to be just as enthusiastic on the last pitch of the day as if it were the first. Because who knows when it's going to be the one that says yes,"

Staying positive and setting the vision is job number one for the CEO, but it's a thankless job as Radford shares, "Don't go looking for 'attaboys' for your good decisions, cause you're not going to get them, therefore be secure in praise solitude. Rarely will anyone say a good job, and frankly you can't rely on those to motivate you or keep you positive. You must be completely self-sufficient. Conversely, people are going to talk shit. Don't take it personally and just let it flow off your back like a duck in water."

Recalculating Customer Lifetime Value on the Farm

For most of the book, I've been coaching you on getting customers, but now that your sales are expanding, it's time to discuss the importance of keeping and growing customers, as Steve Blank taught. In the early days of startups, sales prospecting is like a dart board figuring out anything that sticks. Once you hit a bullseye, it's time to write down that score. In business speak this is determining the customer lifetime value (CLV) or multiplying the average customer sale price time by the lifespan of your retaining them as a customer. These are two variables that are entirely in your control. By increasing spending on your customer success efforts you could also expand the average sale across an organization's business unit, entrenching your products fully within the organization, and as a result lengthening the lifespan of the relationship.

Andersen of Augean Robotics, maker of the Burro autonomous farm vehicles, has institutionalized his account and customer success feedback loops,

FIGURE 8.1 Andersen's newest product Burro Grande at Altman Plants in Giddings, Texas.

Photo credit: Augean Robotics, 2024.

enabling him to introduce a new product earlier this year, the Burro Grande. The Grande can tow five times the weight of its original product, opening up new opportunities for existing customers and new markets. He explained how he captures field data and loops it back into the sales teams, "When I created our customer sales team, we didn't even call it that at that point. It was like we didn't even know what we were creating exactly, but the guys [our employees] who spoke Spanish talked with our customers were also super sharp and were able to relate that feedback to what people wanted. Of any team that we've built out, our customer success function living and breathing in the field has caused us to move so, so much faster than we otherwise would" (Figure 8.1).

Unpacking Andersen's statement, he sells to the farm owners, but the users (mostly Spanish-speaking) are the ones who are most open and knowledgeable about the needs in the field. Building trust with these workers ensures a longer and more fruitful relationship. As he elaborated, "Customer feedback is a ritual, and it's core to what we do every single day. There's a company-wide call; not everyone has to be on it. But as a company, walking through how each of our largest customers has used the product and major feedback loops that we've heard." However, he worries that too many call participants only share what is going well and that there are not enough areas where the company can improve. As he conveyed, "I think what makes a great product is mostly criticism at times. Hey, I don't like it this way. This is rough around the edges. Customer A is going to find that button placement really annoying. How are they going to see that icon? Why does that screen flip or switch when they're going indoors or outdoors or in this context? And so I think that to me the most surprising thing as we've gotten bigger is trying to take really, really talented team members and make sure that they're living and breathing, and how customers are using the product."

The CEO continues to evolve his weekly feedback calls; he maintained, "And so again, we have a daily call every day as a team that is field facing, and then everybody within our team has spent at least a couple of days to a week in the field using the product." From these calls, he turned his attention to Customer Success, as he explained, "And then the final thing I think is a product function for me has been a very tricky thing to build out. People tend to assume that once you have something built, the market will just accept it and move into it. I think there's a big, big process of once you have a technology working, getting it integrated into customer workflows, making sure that it's producing a positive payback, making sure that you know customers are having success with the product. And those kinds of details require a very driven product team and a very driven customer success team that's really capable and you-centric, not robot-centric. People like to talk to robots and lose sight of the users. What matters ultimately is the people using it, getting value out of it, and loving it."

In the case of construction robotics companies like CivRobotics, it's not just one job site but multiple projects across the country that could accelerate the adoption of their system. This corporate expansion provides Civ with numerous opportunities to listen carefully to the customer and upsell new products and services to their internal champions based on continuous feedback. If a customer says for every one of those you sell, I will buy eight of these, then that's your ah-ha moment to land and expand the relationship.

As Tom Yeshurun, founder of CivRobotics, reminded us customer service is vital to expanding relationships from pilots to large recurring revenue streams, "Look, the customer, especially the new ones, they're not going to buy a product early on without trying it for a few months, so call it a rental period. Then there are other ways to keep them happy under the rental. If there's any issue from negligence, whether under warranty or not, we need to fix the machine. You're renting the machine from us, worry-free. If there's an issue, we will give you another one. We'll fly and fix yours on your job site to ensure you are happy and can run your machine with as little downtime as possible. And then, if they really like the machines and have multiple of them in the future, they may want to push us towards purchasing. But that's a different conversation because we're talking volume like we're not selling them one, we'll sell them five. And they are paired with a software subscription and maintenance plan that is again recurring revenues, which is what our investors are interested in, and an upsell opportunity because that machine will not last forever."

Founder Insights: Hailey Nichols, Sensor Fusion

Numerous innovations profiled so far have been complete systems. However, many mechatronic startups are components within larger product offerings, like computer vision for a robot arm or the navigation technology of an autonomous car. These offerings often find sales challenging as their value propositions depend on channel partners, system integrators, or original equipment manufacturers. Consequently, many of these technologies have become attractive acquisition candidates by more significant enterprises, with many founders now having an 'exit mindset' towards selling their business down the line. In reading through Locus Lock's business below, start brainstorming possible strategic partners for the company to hit its next milestone and maybe one day an exit.

Background

The International Air Transport Association (IATA) reported in 2024 an increased level of GPS spoofing and signal jamming since the outbreak of the wars in Ukraine and Israel. The alarming rise of nefarious activity could have catastrophic outcomes for commercial flights everywhere. For example, last September, a European flight en route to Dubai almost entered Iranian airspace (with no clearance).[3] In 2020, Iran shot down an uncleared passenger aircraft that entered its territory.[4] This danger has made the major airlines, avionic manufacturers, and NATO militaries and governments scramble to find solutions.

Locus Lock offers an innovative approach to Global Navigation Satellite System (GNSS) signal processing through advanced software techniques, allowing for high-fidelity positioning at a fraction of the cost of comparable hardware solutions sold by military contractors. In March 2024, I sat down with its founder Hailey Nichols, a former University of Texas researcher in the school's famed Radionavigation Laboratory. Nichols explained her transition from academic to founder: "I was always enthralled with the idea of aerospace and went and studied at MIT, where I was obsessed with the control and robotic side of aerospace. After I graduated, I went and worked at Aurora Flight Sciences, which is a subsidiary of Boeing, and I was a UAV software engineer."

At Aurora, she focused on integrating suites of sensors (lidar, GPS, radar, computer vision, etc.) on autonomous aerial vehicles. However, Nichols quickly became frustrated with the costs and quality of the sensors. "They were physically heavy, power intensive, and it made it quite hard for engineers to integrate… This problem frustrated me so much that I went back to grad school to study it further, and I joined a lab down at the University of Texas," she emphasized. In Austin, the roboticist saw a difference, using software techniques to achieve high-rate signal and data processing. She elaborated, "The Radionavigation Lab was very highly specialized in signal processing, specifically bringing advanced software algorithms and robust estimation techniques forward to sensor technology. This enabled more precise, secure, and reliable data for positioning, navigation, and timing." Her epiphany came when she saw the market demand for the lab's GNSS receiver from the Department of Defense, as well as other commercial partners after Locus Lock's published research on autonomous vehicles accurately navigating urban canyons.

Taking a Software Approach to Hardware

Locus Lock is ready to market its product more widely for dual-use applications across the spectrum of autonomy for commercial and defense use cases. "Current GPS receivers often fail in what's called urban multipath, which is where building interference and shrouding of the sky can cause positioning errors. This can be problematic for autonomous cars, drones, and outdoor robotics that need access to centimeter-level positioning to make safe and informed decisions about where they are on the road or in the sky," stated Nichols. The RF engineer continued, "Our other applicable industry is defense tech. With the rise of the Ukraine conflict and the Israel conflict in the Middle East, we've seen a massive amount of deliberate interference. So bad actors are either spoofing or jamming, causing major outages or disruptions in GPS positioning."

Locus Lock solves this issue by deploying its GPS solution as a software module, and unlike hardware, it's affordable and extremely flexible. As the founder elaborated, "The ability to be backward compatible and future proof where we can constantly update and evolve our GPS processing suite to evolving attack vectors ensures that our customers are given the most cutting edge and up-to-date processing techniques to enable centimeter-level positioning globally." She continued to share Locus Lock's defining advantage over existing products, "Our precision positioning engine is an enhanced form of inertially-aided Real-Time Kinematic (RTK) technology. Essentially, our software uses advanced sensor fusion techniques, combining GNSS signals with inertial navigation to maintain centimeter-level accuracy, even in challenging environments like urban canyons where GPS signals may be lost. In such cases, our GNSS software-defined radio (SDR) helps bridge these gaps to deliver consistent high-precision positioning." Nichols boasted that Locus Lock is an enabler of "next-generation autonomous mobility."

While traditional GPS component manufacturers cost upwards of five thousand to tens of thousands of dollars, Locus Lock is able to deliver its proprietary software solution (with a 2-inch board) for a fraction of the costs (under a thousand dollars). Today, centimeter accuracy is inaccessible to most robot startups as the marketplace for robust navigation solutions is mostly from high-cost hardware manufacturers. In responding to the competition, Nichols stated "We've specifically made sure to cater our solution towards more low-cost environments that can proliferate mass market autonomy and robotics into the ecosystem." She further advanced the flexibility of her GNSS receiver in pulling in data from global and regional satellite constellations. "Giving us more access to any signals in the sky at any given time. Diversity is also increasingly important in next-generation GNSS receivers because it allows the device to evade jammed or afflicted channels," exclaimed the startup founder.

Grand View Research estimates that the current SDR solutions market will climb to nearly $50 billion by 2030. As more uncrewed systems proliferate, Locus Lock's price point will also come down in value. Nichols reflected:

"Although some mobility companies have reached a high level of autonomy in their navigation systems, the costs of commercializing this technology remain too high for widespread adoption. For the next generation of autonomous mobility, access to high-integrity, reliable sensors at an affordable price is essential. Locus Lock is providing high-end performance at a much lower price point." She even predicted that they could eventually get their solution to under a thousand dollars, if not less, with more adoption.

🖎 *Assignment:* Locus Lock is an early startup beginning to engage with prospective customers; how would you organize the sales and product teams to create a robust feedback loop that enables them to capture more revenues by introducing new products from the field? At the same time, how do you make that "exit mindset" in a company spun out of a University lab to begin having strategic conversations with potential partners?

Notes

1 Seuss, Dr. 2016. *Oh, the Places You'll Go!* London Harpercollins.
2 Seuss, Dr. 2016. *Oh, the Places You'll Go!* London Harpercollins.
3 September, OPSGROUP Team 28, and 2023. 2023. "FAA Warning Issued, Further Serious Navigation Failures Reported." International Ops 2024 - OPSGROUP. September 28, 2023. https://ops.group/blog/faa-warning-navigation-failures/.
4 Reuters. 2023. "Iran Taken to World Court over Downing of Passenger Plane," July 5, 2023, sec. World. https://www.reuters.com/world/middle-east/iran-taken-world-court-over-downing-passenger-plane-2023-07-05/.

9

EXIT OPPORTUNITIES

Rappelling

For a backcountry hiker, a summit is the top of a mountain; for an entrepreneur, it is a liquidity event that turns paper stock into cash. This process starts with establishing an exit mindset in building your company with the expressed purpose of going public via an initial public offering (IPO). Everyone invested in the organization is part of this trajectory: the financial backers, strategic partnerships, and employees. Selling one short to build to sell is a recipe for disaster and early shutter. At the same time, if an attractive offer comes knocking on your door, you have a team in place to analyze the opportunity from a point of strength. This chapter will highlight various ways many companies interviewed earlier have exited and returned capital to their initial backers.

iRobot's Initial Public Offering

One of the most prized goals for founders is going public on either the New York Stock Exchange (NYSE) or the National Association of Securities Dealers Automated Quotations (NASDAQ). In fact, in the lounge of the NASDAQ building in Times Square, there is a museum of some of their most valuable holdings, including Apple, Amazon, Tesla, Intel, Airbnb, Monday. com, and Activision. Joining this club is a winner's circle that every founder (and VC) desires. However, the pressure to keep up with its membership with quarterly earnings reports, profit guidance, and analyst coverage is arduous.

Rodney Brooks recalled the rollercoaster of meeting public expectations: "When you do go public it's a whole different set of expectations of meeting what the market expects. And I didn't understand how that was, with the share price just jerking around. But we did really good, yeah, but you didn't. And we did better than we promised, but you didn't do as well as we thought

DOI: 10.1201/9781003508410-10

you would do, so…. we're going to whack you. That to me is what happens." Brooks remembered, "It was unpleasant, having that external force being applied in a way that we didn't fully understand, ever." On the flip side, he shared how the IPO changed everyone's lifestyle for the better, "The IPO experience for us, well, it's changed my life. So it's good in that sense. And a lot of our employees were able to buy very nice houses, which was good. Probably a couple of 100 of them were able to buy nice houses that they wouldn't have otherwise had."

He cautioned founders looking to follow iRobot on the NASDAQ, "This is some advice for the founders. As you get bigger, you're going to come under more scrutiny. Even in Series B there's going to be a lot of scrutiny. Don't do stupid financial things; don't do stupid side deals. 'Cause that, you know, it's going to be a turn-off for every investor, and by the time you get to an IPO it's really going to turn off investors." He summed up that "Everything has to be clean…. So, be clean at all points; it will make everything smoother, including the IPO. You want your underwriters to be saying, we haven't found the skeletons. And if there's no skeletons, you're in a great position." Just in case it is unclear what type of side deals he referred to, Brooks specifically called out, "Don't give yourselves loans. Don't promise this to that person. Don't have all these side agreements that just make things more complicated and hurt the cap table. Resist investors who want to add all sorts of wacko financial terms, because they will hurt you eventually."

Helen Greiner, who became Chairwoman and President of iRobot in 2005 before it went public shared her own experience on the day the Company listed on the exchange. Leading up to that moment, iRobot had already sold one million Roombas, which has climbed to over 40 million today. She recalled, "It was very dicey for a while. But everything started going up when Pepsi put out a commercial with a Roomba on it. And then we raised more money. We didn't raise a round based on Roomba until after it was on the market and successful. And even then I'll tell you it was tough, even after we had a plan to go public. It was all looking promising, but we needed money." To fill that immediate cash crunch, Jeff Bezos (founder of Amazon) came to the rescue as a long-time supporter of robot ventures. Greiner recollected the IPO experience, "I could say we had a little bit of a hiccup when we went public in 2005. We went on the market, I think at 24 [dollars a share] and it had gone up by the end of the day. Then that night the guy from Mad Money put us on his show. He raved about it." However, as she learned Jim Cramer's upbeat report was both extremely positive, and, unfortunately, a curse as she recalled, "And it went up, and all the great work we did getting long-term investors with the roadshow. We bet our target it's like 38 [dollars a share] or whatever. They sold, and so then we had a lot of retail investors who were just playing momentum games with the stock. The investment bankers took us aside after the roadshow, and said, 'This guy didn't do you any favors. This is not positive. It looks good now, but it's not a positive thing for the long term. People going to dump their stocks and people will short it as people are dumping.'" This hard first lesson of being a publicly traded stock stayed with Greiner almost 20 years later (Figure 9.1).

She also thinks it's crucial for companies not to lose their innovative spirit while scaling to meet the market demands of a listed entity. Almost twenty years later, she stated, "I don't think it changed too dramatically. You know I'd dress nicely everyday instead of like now. There was a lot of emphasis on getting the quarters right, which is not that positive. But I really feel strongly. It was the right decision and the right timing for us. We owed it to the investors who had been with us since '98, and the people who've been at the company since the beginning, like myself, to make the stock liquid in some capacity. So, I made a deal that everybody could sell the same percentage of shares at the IPO." She summed it up, "So it changed our lives, meaning that we for the first time had some money. But the company, I think, was the same culture we had for the first few years, except we had more things we had to do."

FIGURE 9.1   iRobot's founding team: Colin Angle, Helen Greiner, and Rodney Brooks (with Roomba and PackBot on hand).

Nauticus Robotics: Understanding SPAC Math

Most publicly listed stocks, like iRobot, go public through traditional means of filing a new listing on the exchange with underwriters who draft a prospectus and S-1 filing with the Security and Exchange Commission. Today, this process can cost a company upwards of $10 million in underwriter fees, lawyers, audits, accountants, and other consultants. There are different ways of getting listed on the exchanges, such as a direct listing (without underwriters) or a reverse merger of a private company to a public 'shell' that has a ticker but no real assets. The latter is called a special-purpose acquisition corporation (SPAC), whereby the public shell raises capital to acquire a private company.

Until May 2021, SPACs were extremely popular for venture-backed startups to raise cash and as an accessible means to obtain liquidity for their shareholders. Many of these stocks today have lost over 90% of their IPO value due to a lack of institutional following and low market volume. These two elements are critical, you need analyst coverage to promote the stock to market makers, and enough buyers and sellers that want the stock to be able to trade. If there is little float/volume then one person with a lot of shares selling could dump the stock and it will fall swiftly. This risk is compounded by the added costs and pressure of being a publicly traded company that has to report quarterly. I've personally experienced holdings that went public via SPAC with great enthusiasm at the opening bell to crashing a year later with the underwriters running to the bank like Mr. Monopoly with their fees in their pockets.

Nic Radford shared his experiences in taking Nauticus public via SPAC as a warning to future founders looking for cash or liquidity, "If anyone ever came to me today and said, 'Hey, we've got this idea, we've got a SPAC interested in X, Y, or Z, and we'd like to talk to you.' I wouldn't even pick up the phone." The former CEO of a former high-rolling SPAC merger opined, "The whole SPAC industry seems to be very misleading, and it's really unfortunate the good and promising companies that they exploit. If you really think that your company has the profile or the fundamentals to be a public company, there's no reason on God's green earth you would do a SPAC outside of traditional S1 offering and doing an IPO."

He further pulled the curtain on what underwriters/bankers falsely promoted, "The allure is since it's a merger and it's only an S4 registration statement." They sell you on that it's going to be faster and "more cost-effective to do," which might be true under certain circumstances, but for the majority it turns out to be misleading at best. I'm sure there are exceptions, but the deal costs and the complexities of doing an S4 merger are crazy and there's so many hidden aspects to it." Radford did not hold back and now considers the industry full of half truths. "The intoxicating part is they flash this huge trust in front of you, and SPACs range from $50 million trusts to half a billion dollar trusts, and you're looking to the next level to fund a growth plan, and they say, 'Hey, look, here's a path to have liquidity. Here's a path to put a bunch of cash on your bottom line, and then you can go execute that business plan.' And it all sounds amazing. You're like, oh my goodness, and the lock-up periods aren't even that bad. Typical lock-up periods for the incumbent, on the company side, are six months. The SPAC insiders even say we'll lock ourselves up for 12 months. But there are just so many nuances that they don't even tell you about," the former Nauticus CEO reflected. He continued to list, "Not to mention executing a volatile business plan under the guise and scrutiny of an unforgiving public investor."

He cautioned founders to read the details, "First of all, fees, the fees in a SPAC are metaphorically criminal. And I have no trouble going on record saying any of this stuff, the fees in SPACs are enormous. The SPAC sponsors and the folks that organize the SPAC typically take 6% to 8% of the company as founder shares. So they say we will take a certain percentage of the company as an enterprise value, plus the cash in the SPAC. They set that and say we will take 6% value. The problem is, by the time you de-SPAC most of the trust money goes away."

As he elaborated on the underwriters' hidden math, "So let's say you have an enterprise value of 100 million, and there's 100 million in the trust, so you've got a $200 million deal, they're going to take 6% of that deal, right? So they're going to take 6% of 200 million. But at the time you de-SPAC [publicly list], when I was doing it, everybody was seeing 95% redemptions. So of the 95% redemption, you get $5 million out of the trust. So now you have a $100 million company, plus $5 million in cash, and they still take the 6% of $200 million of your company. So that's just the sponsors, right? It's utterly nuts that they take so much of a fee for just this, for what they call founder shares."

He said that was just the beginning of the tally, "Then, you have the bankers. The bankers come in, and they're taking their own cut. Let me just tell you that we had a trust of $175 million and we got $15 million out of it, $15 million out of trust! We raised about $69 million in the PIPE [Private Investment in Public Equity]. We were to pay our banker $12 million to do the deal - cash. We settled for $7.6 for the sell side banker - all cash. And then you have to go find two to three capital market banks to drive interest, all wanting heavy fees. And here's the other thing they don't tell you. So the SPAC goes through an IPO right before they've ever met you and the SPAC IPO puts the trust together. However, some defer that fee, and you could end up paying it when you de-SPAC. So now you may have to pay the IPO fee on the trust, you pay the buy-side banker's fee. A couple of points: you might pay the sell-side bankers fee, and then you pay all the lawyers involved."

As he rapped up, "All in, our deal costs were on the order of $20 million on the SPAC, the PIPE that we raised plus the trust, we had $82 million, and we paid $23 million in fees! So basically, that was an unbelievably expensive Series C round or Series B round, right? I mean, it's like you're paying 30% to just go public." Besides the fees, he decries the realization, "There's no liquidity in the stock. No one's trading it. And even when you're unlocked, you can't sell anything. And as the CEO, don't even think of selling - they warn you."

This horror story was compounded by seeing how all the service vendors benefitted, "And then the SPAC sponsors and the people that were associated with it, they've got zero cost basis, so the second they're unlocked, they just sell because it was free money. And then there's where the stock rights and warrants are trading, which puts a downward pressure on price. And so every SPAC seems doomed from the beginning. With no liquidity, there's no trading volume, there's no float because you took all the money out of the trust, you paid everybody out of the PIPE that you raised, and then you're locked up. But it doesn't even matter, because you couldn't sell it anyway, because there's no one there to buy it. And then when everyone becomes unlocked, you still have no buyers. Everyone's selling and everything drives [down] to $1."

Lessons Learned

When asking him what he would've done differently, knowing everyone's hidden fee now, Radford suggested, "I would say the SPAC needs to pay all the fees or at least guarantee a minimum from the trust. In the later years of the SPAC craze, hedge funds were just using the SPAC trusts to earn a fixed income coupon rate. They get a free look at a company and then redeem earning a great yearly adjusted return. That's a misaligned incentive. I would also cap everyone's fees. Say, all right, you guys want to do this job? We're going to cap all legal at a million dollars. You're just using template forms anyway. We're going to cap all bankers at a million dollars, and if you're still interested in doing this deal, then fine. Otherwise, this deal isn't for you. I would set it in very strict caps. I would also negotiate heavily down the founder shares, which is essentially a commission that they get for finding a deal. I'd say no, you guys aren't getting 8% of the trust plus the company's enterprise value. You're getting 3% of whatever the final deal value is." He also recommended, "I would have said I want a minimum cash on the balance sheet of $125 million. If I don't get that, we're not closing the deal and everybody's not getting paid. All the legal work should be on contingency of closing the deal."

He further advised against a SPAC entirely and that founders go through the traditional IPO process if they must go public, "Because with a traditional IPO, you have one registration statement, you have an S1 registration statement. With a SPAC, you have not only an S4 merger registration statement, but then you still have to file an S1 anyway. So it's not like you're saving anything. You just get to lengthen the process. So maybe you could probably get to a public-facing company sooner, but you still have to file an S1 and then they don't tell you all the legal expenses. Do you know how much money it takes to run a public company? It was running us about $6-8 million a year to run a very small public company." He broke down his costs, "An audit firm to do our quarterly audits was almost $300-400K every time, every 90 days. And that was a cheaper one, and we had a company doing $10 million in revenue. Our Directors and Officers [D&O] insurance was about a $3 million year premium for the first year, because SPACs have a higher risk profile, so they charge insane premiums. We spent $3 million on ongoing legal, our first year. It was absolutely insane, the amount of costs involved, and nobody tells you any of it."

Finally, he outlined how it changed his job and the entire culture of the firm to be a public company, "So from a workload perspective, it's crushing because, at a small company, everyone's doing four or five jobs. So personally, I'm handling investor analysts. I'm handling inbounds from large shareholders. I would get LinkedIn messages from large shareholders in the open market requesting a meeting because they have a large block of shares and want to talk to the CEO. So, from a workload perspective, it takes enormous energy just to keep the machinery running for a public company. There are so many statements that I have to attest to as the CEO alongside the CFO." This was compounded on top of the core responsibilities of a CEO in running a fast-paced, highly technical robotics company, "You've got a whole bunch of folks that surround a company that you have to attend to at the same time you're trying to manage the internal aspects of the company. So you're trying to handle your C suite and all of the technical challenges and make sure that the company's marching towards the right objectives. You're then managing the board and your staff."

He remembered the craziness, "Just do this one thing the public wants to see, the euphoria of we're going to take over the world, and your company is yelling at you that everything is gonna be very challenging and we're not going to meet our milestone because of unforeseen delays. For us, we had just come out of COVID and you couldn't buy anything, so everything was massively delayed. So it's just this pressure cooker from three different sides."

In conclusion, he warned founders against the seduction of money and the glitz without first calculating the costs and risks: "When we were going public, there's a lot of pomp and circumstance. You're on CNBC. You have several interviews with notable outlets. I was at the NASDAQ. I gave a speech at the bell. Everyone comes to see your picture on the seven-story tower,

NASDAQ Tower, in Times Square. There's a lot of really good energy around it, and you feel like, wow, we are doing something big. Maybe this thing is going to work out. We made it. Now here is the travesty of it, all the people who have come into the company now and have received millions of dollars of shares at 20 cents are likely going to make a killing over the next year, but a lot of all the original folks are gone, so the company is sort of turned over. It's got a whole bunch of names I don't even recognize at a company that I started in my living room, and they're all going to make money and a lot of the original founders have been pushed out."

Corporate M&A: Case Studies

Most startup exits are acquisitions led by larger organizations (aka Big Corporate). It is important to note that this activity is affected by interest rates and as of the writing of this book, 2024 has been one of the lowest points in almost a decade for corporate mergers and acquisitions and public offerings. While there has been a smattering of successful IPOs in the mechatronics and hardware space, a greater number of private companies have sold to synergistic conglomerates to dominate their industry successfully. For example, humanoid and anthropomorphic robotics leader Boston Dynamics has gone through consecutive owners with Google, Softbank, and Hyundai. The capital and market distribution that a good partnership brings could be a game changer in taking technology from a niche project to mass adoption. At the same time, merging a startup culture with an international organization is easier said than done. To navigate this path, I have summarized some ideas from our sherpas, Ryan Gariepy of Clearpath (sold to Rockwell Automation in 2023) and Mick Mountz of Kiva (acquired by Amazon in 2012).

Clearpath Sells to Rockwell Automation

There is no such thing as serendipitously regarding acquisitions; strategic discussions start with what value a startup could bring for a Big Corporate acquirer (i.e., sales, technology, talent, etc.). You have to be attractive enough just to begin a conversation. Gariepy stated this process takes an investment of time, "This is why it's important to always be out there talking to each other, talking to the world, going to the right trade shows, and having these conversations. I think our first contact with Rockwell was before COVID, 2019 or earlier when we started being aware of what each was doing in the AMR space." He continued to share how it evolved into more of a strategic relationship, "Over time, we were talking more, having conversations about where we can help each other out. Where can we grow the collective pie for everybody? Going into that without an ego, knowing that we can help these large companies, but that these companies can help us out too. Likewise, I think they looked at what we had and saw value there. Obviously, in the end, we know what happened." More specifically, he said, "We're coming into fundraising, our Series D, we retained a banker, and we also had some conversations as a board and founders about how we felt about selling. As the founders, we have been at it for 15 years, which is a long time. We also saw a number of other market dynamics occurring. There were statistically very few robotics companies, if any, which had sold for north of 1 billion dollars."

He broke down how the math of their Series D supported the rationale for a strategic sale, "We knew that we were raising money on certain terms, where we would be at a 700-800 million dollar post [money value]. This means then that you need to sell for north of 1 billion dollars, and that means IPO, because to do M&A at this time in robotics, you enumerate on one hand the companies around the world that would pay in that range. We also knew that the presently interested parties weren't likely to be interested in the future, because they wanted to make a move quickly. If we didn't sell, we would actually end up competing with the other parties in the future."

He succinctly summed up his thinking, "You get a bit of a power law where if you're valued at a million dollars, there's 100,000 companies out there that might buy you. But as you get to a billion dollars, it was 'IPO or bust'. We also continue to really want to make an impact on the world. And we saw Rockwell as a team that would support us in doing so."

Founder Calculus

Gariepy opened up to me to share his thought process on the transaction, which offers every founder a perspective for making such an enormous leap. "I really like technology, and we know that as you scale as a company, more and more operational stuff comes into play. And that's the other thing founders have to ask themselves: 'What motivates you? What do you like waking up and doing?' As you become a larger and larger company, there's more operational stuff." he reinforced.

In looking at what Rockwell had to offer, the CTO elaborated, "We can all have this giant impact, and that's was a really exciting idea for us. I'd say I personally wanted to grow and learn as a technologist and as an engineer." He noted that "Bringing in funding from a PE [Private Equity] firm or doing an IPO wasn't going to necessarily wind up being a good outcome for me as a person." On the pro side, he elaborated that "Joining a company like Rockwell, which has thousands of really excellent engineers and technology leaders gives me the opportunity to learn every day." It's vital, said the founder, that every entrepreneur asks themselves what they want to do for the next ten years. For Gariepy, reporting to a PE buyer or the public markets was less exciting than working with talented like-minded innovators. Most importantly, he coached, "Don't get greedy and overfocus on maximizing your personal financial outcome. This was a life-changing event not only for myself but many, many people on the team. You only live once."

Operational Steps

Once you decide to move forward, it's time to put the wheels in motion. Note that once the car is moving downhill, the inertia of the transaction is hard to stop, so decisiveness is essential to closing the deal. In listing advisors that helped Clearpath close, Gariepy said, "I would say that obviously having a good banker is great." However, even with great bankers the devil is in the details, and while Rockwell's acquisition of Clearpath was relatively quick, it was a lot of work. He reflected, "Our due diligence was three months of every question possible. I don't know how many thousands of pages went back and forth, but because we had a great operational team, I could focus on the product and technology story, Matt could focus on the sales and the overall vision, and we also could keep the business growing. And when it came down to it, it was important to get it done quickly because time kills all deals."

Similar to Brooks' comments about having a tidy house, Gariepy stated, "The biggest thing I would say it's about the decisions we made years earlier, always try to make sure you've got a clean cap table, clean governance, that you don't have complicated pref stacks and things like that." Clearpath's attention to keeping each prior financing 'clean' without preferential treatment to different shareholders and with proper data rooms aided in accelerating the deal. Founders need to recognize that it is in their best interest to close quickly because every delay could risk the deal with so much money and time already invested in the transaction. He shared that he was worried about their business model holding things up, "We did not have a simple business model at the time. There's going to be risk there. It will never be entirely simple. There's always going to be warts and nuances and things like that. But the simpler you can have it. The better." He credits his co-founders' integrity and transparency as the reason why they were able to complete such a significant transaction. "I think those decisions we made years earlier, sometimes over ten years earlier than the acquisition paid off."

The other important piece of advice he has for founders is to be careful of what you wish for in terms of valuation. "I know that there's a lot of pressure as founders to just chase the numbers. We never optimized for valuation, and that was very good for us. We had a very good outcome. As far as I'm aware it was the third largest mobile robotics exit in history. That's an excellent outcome," counseled Gariepy. He thought it could have killed the deal if they had been more aggressive early in driving up the valuation, "If we really tried to optimize our valuation, as some companies I see are doing, I think it would have been more challenging. We may have had buyers just not engage because they know what our valuation was, and they'd assume we're not gonna take their offer."

Internally managing shareholder expectations for liquidity is critical, and high valuations could also lead to shareholder disputes if the market corrects. Instead, he related that, "All of our shareholders said, 'This is a great deal. We support the recommendation to sell, and that was very positive. Again, I think a lot happened because we didn't get greedy. We didn't try to optimize for valuation. We tried to be reasonable when we were raising capital." He worried that today, there are hyperbolic valuations that are not substantiated by financials or sales projections. "You see some robotics companies that have a series A valuation in the range of 1 billion dollars. Are they going to sell their Series D for 5 billion dollars? I think it's not worth the roll of the dice given the present maturity of the robotics field," claimed Gariepy.

Amazon Bought Kiva, the Sale That Launched a Thousand Startups

In 2012, robotics as an investment class was very nascent. If today we are in the third inning of the automation series, when Amazon acquired Kiva Systems for $775 million, we just took to the field. Following that Kiva acquisition, Google also purchased seven robot startups, but that was Larry and Sergei's moonshots with Andy Rubin, which operated outside the world of private capital. Kiva on the other hand was the catalyst for showing the industry that there actually was a pot of gold at the end of the mechatronic rainbow.

Raffaello D'Andrea looked back to that time and remembered his fellow founders were not in the market for selling the business, "So first of all, the founders didn't necessarily want to sell. The Board felt that it was a good offer. It was eight times revenue, which at the time was a really good multiple." He also credited the market cycles, falling a few years after The Great Recession, "I guess maybe they got spooked from 2008. And they just wanted to have a good outcome." Asking him today as he runs his next startup, Verity, he noted, "In hindsight, you could view it in two ways. One is about five years after the acquisition. There were some estimates of how much money our system was saving Amazon. And it was something like $2.5 billion a year. So basically, the payback for them after 5 years was like 3 months.

And I'm guessing that now, 10 years after or 12 years after, it is probably close to $10 billion." When I asked D'Andrea about seller's remorse, he replied, "That's neither here nor there. You also have to say that Amazon took what we created, and they built all of Amazon's robotics on top of that. So for me, it was a great outcome."

The first thing that Amazon did after acquiring Kiva was to make the technology exclusive to its own fulfillment centers, but this was not always the intent of Mick Mountz when approaching the e-commerce behemoth. As Mountz recalled, "We had about 12% of the top 100 e-commerce companies using Kiva in 2011, and at that point, Zappos was our customer and Diapers.com was a customer of ours." He provided a little history of their conversations with the e-tailer, similar to Clearpath, which is about maintaining the relationship. As Mountz explained, "In 2009, Amazon bought Zappos and got their first look at the Kiva System up close. We tried to leverage that into a project with Amazon but couldn't get anything going. Then in 2010, they bought Diapers.com, and the entire Diapers.com system, in all three warehouses, was 100 percent Kiva." After Diapers became part of Amazon, Mountz returned to Seattle in 2011 and said, "Let's do something together, and that's when it turned into a Kiva acquisition conversation."

FIGURE 9.2 Raffaello D'Andrea demonstrating Verity drones at the Unexpected Sources of Inspiration conference.

Photo Credit: USI 2019.

The acquisition of Kiva is a symbol of the fruition of all the ideas expressed in this book, which take one's passion for disrupting and improving work-flows to becoming the de facto platform that automates everyone's lives. Mountz experienced firsthand a problem back in the dot com bubble with Webvan. In the early 2000s, he organized his solution into a business that not only tackled the original problem but accelerated the entire e-commerce industry with streamlined fulfillment. Now today, Mountz's vision touches everything inside that smiling box delivered to your door. To put Kiva's impact in perspective, there are nearly 10 billion packages shipped each year worldwide (Figure 9.2).

Other Exit Opportunities

Going public or selling the company is one of many ways for a founder and their shareholders to convert their equity into dollars. Depending on your sales and profit margins and synergies with other holdings, private equity (PE) firms are actively looking to roll up businesses within a particular industry. In addition, these PE managers also like to marry established borning cash-led manufacturing concerns with exciting technology like automation. Typically, a PE deal requires sales north of $10mm and with 20% profit margins. The sale structures of these transactions vary, but PE owners set out to acquire the majority of the company's assets and keep management on as an earn-out (whereby part of the deal is paid upfront, and the rest is milestone-based over time). From my experience, earn-outs seldom achieve the desired objectives for both parties and provide a way out of the obligations for an acquirer. Also, for many founders, shifting from boss to employee can be unnerving and lead to conflicts with your new supervisors who are now in charge of the vision and rearing of your business child. To maintain control while taking cash off the table, you may consider a growth equity buyout, where a later-stage venture capital firm would look to buy shares from your equity holders and employees for a minority interest. This transaction takes on different forms (both primary and secondary purchases of stock, and debt), and like all the transactions above it is best run through an organized process with a leading banker in your industry. Finally, there are Secondary buyers of stock, whereby your shareholders and employees can sell their equity through a broker, which is less involved and often at a steep discount. The point is there are options, and if your goal after five or ten years is to buy a new house and continue to run your company, a bird in the hand could be worth two in the bush.

Founder Insights: Robot Venture Factory

People still have post-traumatic stress disorder (PTSD) related to COVID-19. Today, in New York City, there are still people walking around with double masks outdoors. However, most urbanites have returned to normal. In many ways, robots had a black swan moment during the pandemic, especially cleaning bots that emitted blue ultraviolet light to disinfect hospital rooms and nursing centers.

Background

In March 2020, New York announced that it was calling 40,000 formerly retired healthcare professionals to augment its current hospital staff.[1] Administrators in other places, such as China and South Korea, deployed robots for sanitation, deliveries, and even, patient examinations. A month prior, *The Robot Report* published that Blue Ocean Robotics had shipped 2,000 units of UVD Robots, a mobile ultraviolet sanitizing platform, to China to combat COVID-19. In March 2020, I caught up with Claus Risager, founder of Blue Ocean Robotics, to discuss how the global pandemic has impacted their business. The Danish roboticist has been at the forefront of robotics innovation since the late 1980s. *Note that the opinions of Risager expressed herein are his own as co-founder and chief executive of Blue Ocean Robotics in 2020, and they do not represent the views of the current management team.*

In 2013, Risager set out "To develop mobile robots that are tailor-made to specific service applications in healthcare, hospitality, construction and agriculture." As an anecdote, I first met the Blue Ocean Robotics team when they approached my portfolio company, Que Innovations, for autism therapy devices. Since then, I have seen the Nordic startup lead in shaping robotics worldwide. As the founder described, "The company is a 'robot venture factory' in the sense that we have a portfolio of robots each marketed under its unique brand name. Today we have UVD Robots, GoBe/Beam Robots, and PTR Robots in our portfolio." Risager also promoted "robotic entrepreneurship" by cycling best practices and reusing components to commercialize new innovations. As the CEO explained, "We have been able to streamline the process of going from an idea, through design, development, and commercialization all the way to scale up and doing so in a way which is better, faster, and more cost-effective than others due to the reuse of well-proven components. Thus the capital it takes, the risks, and the time it takes are significantly reduced. Today our robots are sold in approximately 50 countries around the world and we are present in the Americas, Europe, and Asian-Pacific regions."

Healthcare Business Explosion

In 2019, Blue Ocean Robotics famously acquired the assets of telepresence leader Beam from Suitable Technologies in the United States. This game-changing deal expanded Risager's market share in the remote office and tele-health space, especially during the present health crisis. As of March 18, 58% of America's knowledge workers were using online video tools and telepresence applications to maintain employee productivity. In contrast, pre-virus, in 2018 only 5.3% of Americans worked from home. Analysts predict that the current uptick will spill over to the robot telepresence market that was previously estimated (pre-pandemic) to grow to over $300 million by 2023 (based upon post-COVID adoption, it could be as much as 10 times the original projections). Risager succinctly summed up Beam's new business, "In short the telepresence robots change the way business is done and how people interact in general."

Regarding disinfecting robots, the inventor elaborated, "Over the last six years we have been working with bacteriological and virological specialists from local Danish hospitals to make a mobile robot which can assist cleaning staff with disinfecting patient rooms." He expounded on the impact vis-a-vis COVID-19, "We introduced the UVD robots into the market in 2018 and have seen significant growth rates of 400% annually. But clearly, with the coronavirus crisis, there is a very high demand for our UVD robots and we are currently doing everything we can to produce and deliver all those that are ordered from all over the world. We are proud that our robots are directly in action in places like Wuhan in China, Italy, South Korea, Taiwan, Germany, USA, and many other places as well." The results have been promising as it reduces cleaning supplies, materials, and time spent by employees. "Various test results indicate that we are able to reduce the number of infected people by 40-50% in those hospital rooms where the UVD robots were applied," shared Risager.

Blue Ocean Robotics sees UVD's sales growing substantially, from medical facilities to cruise ships, transportation hubs, retail stores, and nursing homes. In 2018, Grand View Research already reported that the market for "global antiseptics and disinfectants" was over $16 billion. However, this does not account for the post-coronavirus effect or the use of machines. Risager observed, "In the pre-corona period we had experienced a solid growth in both telepresence and UVD robots. This shows that Corona or not there is a significant market potential and pull from the markets towards such types of robots. After the Corona crisis and with contingency plans I believe everyone wants to have better tools in their hands to deal with such kinds of crises. Another thing we also see is that people very quickly get used to working in different ways – more digitally – and this I believe will be a major driver forward for especially the telepresence robots. For the UVD robots, the fact that the market is no longer just hospitals but also the other market verticals will mean that the need for these robots will most likely continue to be very significant."

Sitting in America's worst-hit metropolis, I pressed Risager for his opinion on the new normal. "First of all, I believe the post-corona period will be one where every country on earth will start working more on establishing contingency plans for pandemics. Due to SARS and MERS countries like Taiwan, Singapore, and South Korea already had very fast responses and very detailed contingency plans for virus outbreaks, and thus, these countries do not have exponential spreading of the virus. They have a constant number of people being infected and this completely changes the situation such that closing down the society and business is not necessary," opined the robot executive. He remarked further on how his business is responding, "The coming year seems to be very much focused on a further scale-up of our sales of telepresence- and UVD robots. More and more solutions providers/distributors are on-boarded, and more and more customers are starting to use our robots or increasing their fleet of installed robots." His biggest challenge right now is scaling up. Still, he optimistically stated, "We have also moved to a new and much larger building which enables us to produce thousands of our robots every year under the same roof and where we also have outstanding facilities for our development and testing."

Assignment: Blue Ocean Robotics remains at the forefront of healthcare innovation. After analyzing the interview at the height of the pandemic, what strategic options would you advise Risager to use to grow his business? As he seeks future investment in manufacturing facilities to fulfill orders, could this be the right time for an IPO or PE sale, or is it better to stay private? You are now Blue Ocean Robotics' banker, put together your market assessment and list of suggested exit opportunities

Note

1 "40,000 Health Professionals Sign up to Volunteer as Part of NY's Surge Healthcare Force." 2020. NBC New York. March 25, 2020. https://www. nbcnewyork.com/news/coronavirus/40000-health-professionals-sign-up-to-volunteer-as-part-of-nys-surge-healthcare-force/2343909/.

10

BE HUMBLE AND CAREFUL OUT THERE

Field Notes

We are all familiar with the war motto, "No man left behind." Battles with guns and bullets are often lethal. The wind-up of a startup could be more deadly, at least reputationally speaking. It is easy for operators to develop (and buy into) the hubris of their idea and its success. Confidence and optimism are part of being a visionary, and arrogance is just being a jerk. This conceit leads founders to ignore their early backers, often family and friends, who believed in the idea because it was you. Don't fall off this cliff, as getting back up requires a helping hand. The startup community is too small, and poisonous snake oil salesmen are plagues that need to be avoided.

If you are reading this saying, "That's not me," remember that the trail is lengthy, covering possibly over a decade of your life. At different junctures, you will be faced with ethical and business challenges; how you address this will determine who will join you on your next adventure. The final analysis starts with a great solution to a mission-critical problem that people will pay lots of money to solve. After finding product market fit and a reliable and affordable system, scaling up is easy, well, at least manageable. While many quote the law of averages to discourage founders from pursuing their dreams, I prefer the Doctor:

> Today is your day!
> Your mountain is waiting.
> So...get on your way![1]

Note

1 Seuss, Dr. 2016. *Oh, the Places You'll Go!* London HarperCollins.

DOI: 10.1201/9781003508410-11

APPENDIX

Lexicon of Startup, Robot, + AI Terms

409a Valuation	An independent and legally required appraisal of the fair market value (FMV) of a startup's common stock, reserved for founders and employees, typically marked at a fraction of the price per share from the last financing.
83(B) Election	A U.S. tax provision that enables a founder or an employee the option to pay taxes on the total fair market value of restricted stock at the time of granting versus vesting.
Accelerator	A mentor-based program that propels a startup's launch, market-entry, and funding through hands-on support, guidance, and structure over a period of time (e.g., 3 to 6 months) that culminates with a public Demo Day.
Accelerometer	A sensor that measures acceleration forces is often deployed to estimate in which direction a robot is moving or which direction is 'down.'
Accredited Investor	A legal income and net worth threshold for US investors, as the law states that only people who are accredited can invest in a startup under RegD, rule 506(c).
Acquihire	Typically, a below-market acquisition of a startup based upon its talent pool vs. its product or service, what some investors would call a dignity exit (no upside, but bragging rights).
Actuator	A device that makes a robot move, typically utilizing electronics, hydraulics, or pneumatics.
Agile	A collaborative approach where teams plan, execute, and improve continuously with input from stakeholders at every stage.

AI Plugins	AI plugins are software components that allow AI systems to integrate with external applications and services.
AI Search	AI search lets users find information through natural language queries instead of keywords.
Allocation	Allocation refers to the portion of a funding round reserved for a specific investor or group of investors, typically communicated as a dollar amount.
Alpha Version	Alpha is a pre-release version of a product, typically still in the testing phase.
Alpha Returns	Alpha returns refer to the excess gains a VC fund generates beyond the market average, reflecting the fund's ability to outperform through successful investments.
Angel Group	Angel group investors are networks of individual investors who pool their resources to invest in early-stage startups, often providing capital, mentorship, and connections.
Angel Round	An angel round is a funding stage where angel investors contribute early capital, typically their own, often through convertible debt or SAFE notes, preceding venture capital involvement.
Animatronic	A robotic system mimicking a living creature.
Annotation	Annotation is the process of enriching data with additional labels, enabling machine learning algorithms to better understand and learn from it.
Annual Recurring Revenue	Annual Recurring Revenue (ARR) is yearly income from subscriptions, reflecting a company's growth and financial stability.
Anthropomorphism	Anthropomorphism in robotics is how humans attribute human-like traits or behaviors to robots, making them easier to relate to and interact with.
Anti-Dilution Clause	An anti-dilution clause protects early investors, such as VCs, from losing equity when a startup issues new shares at a lower valuation, adjusting their ownership through mechanisms like full-ratchet or weighted average.
Arm	A robot arm is an interconnected series of rigid links and actuated joints designed to manipulate an end-effector or tool in space, providing precise movement and support without including the end-effector.
Articles of Incorporation	Articles of Incorporation are legal documents that register a startup as a corporation, outlining its name, purpose, address, share structure, and board details, establishing it as a separate legal entity with liability protection and the ability to raise capital through shares.

Articulated Manipulator	An articulated manipulator is a robotic arm with segmented links and joints, each providing a degree of freedom for precise and flexible rotational and translational movement.
Articulation	Articulation refers to a jointed structure, like a robotic manipulator, with joints that enable vertical rotation and horizontal elevation, allowing access to confined spaces.
AGI	Artificial General Intelligence (AGI) is AI that can learn, reason, and adapt across diverse tasks, mirroring human intelligence.
Associative Memory	Associative memory enables a system to store, retrieve, and process information by recognizing connections, allowing quick access to relevant data for decision-making.
ASR	Automatic Speech Recognition (ASR) is a technology that converts spoken language into text.
AUV	An Autonomous Underwater Vehicle (AUV) is a torpedo-shaped robot used for scientific exploration.
Axis	An axis refers to a specific direction that defines a robot's movement, either in a straight line (linear) or rotationally (rotary), enabling precise control over its motions, such as reaching, lifting, or turning.
Backlash	Backlash is the play in a gearing system caused by imperfect gear alignment, aiding lubricant circulation but potentially causing vibrations and reducing precision.
Balance Sheet	A balance sheet is a financial statement that summarizes a company's assets, liabilities, and shareholder equity, providing a snapshot of its financial position at a specific point in time.
Base	The base, or base link, is the stable platform of a robot arm that supports the first joint, providing a stationary foundation for the entire system.
Benchmarking	Benchmarking evaluates performance by comparing financial metrics or product features to industry standards or competitors to identify improvements.
Beta	Beta is a pre-launch product testing phase, following alpha, where startups use private (invite-only) or public (open) releases to gather feedback and fix bugs.
Bio Inspired	Bio robotics involves designing robots inspired by living organisms, merging biology and robotics to create bio-inspired systems that emulate natural mechanisms.
Blended Preferences	Blended Preferences refer to a situation where all classes of preferred stock share equal rights to payments during a liquidation event, without any class having priority over another.

BMI	Brain-Machine Interface (BMI), also known as a Brain-Computer Interface (BCI), uses invasive or noninvasive sensors to translate brain activity into commands that control robotic systems, such as prosthetics or autonomous machines.
Board of Directors	A company's board oversees strategy, finances, and leadership, with lead VCs often securing a seat to protect their investment; an odd-numbered board prevents deadlocks, while experienced members provide crucial guidance and accountability for long-term success.
Bootstrapping	Bootstrapping is the process of launching a startup using personal savings and early revenue without external investment, allowing founders to retain control and prove viability before seeking funding, after which the company is no longer considered bootstrapped.
Bridge Loan Financing	A short-term loan provided to a startup by investors to sustain operations until the next funding round.
Burn Rate	The speed at which a company expends cash to cover operating costs, typically measured monthly or weekly. Managing burn rate is vital for extending runway and minimizing the need to raise capital under adverse conditions.
Buyout	A buyout is an investment where an investor gains control or a stake in a company by combining primary shares (new capital) and secondary shares (ownership transfer). In a minority buyout, less than 50% of shares are acquired, retaining owner control. A growth equity buyout funds expansion while founders keep strategic control.
CAC	Customer Acquisition Cost (CAC) represents the total expense incurred to attract a new customer, encompassing sales and marketing efforts. It serves as a key metric for gauging acquisition efficiency, often measured against Customer Lifetime Value (LTV), with an ideal LTV/CAC ratio of 3:1 to ensure sustainable growth.
Cap Table	A cap table, typically in spreadsheet form, outlines a company's equity structure, detailing all securities—such as shares, options, SAFEs, and notes—while showcasing each investor's holdings, their value, and ownership percentage.
Capex Sale	A CapEx sale in robotics involves selling long-term robotic equipment or systems to free up capital, reduce operational costs, or upgrade to more advanced technologies.
Carousel	A carousel is a rotating platform that serves as an object queuing system, delivering items to a robot's loading/unloading station.

Cartesian Robot	A robot that operates along a Cartesian plane, moving in straight lines at right angles. Its straightforward design makes it easy to build and program, particularly for large-scale applications. Also known as a linear robot.
Cash Flow Positive	A company is cash flow positive when it generates more cash than it spends, signaling an increase in liquid assets.
Cash Flow Statement	A cash flow statement is a key financial document that tracks a company's cash flow and, alongside the balance sheet and P&L, helps manage finances.
Centrifugal Force	When a body rotates around an axis away from its center of mass, it produces an outward radial force. To maintain control, the robot must apply opposing torque at the rotation joint.
Chasm	The chasm, introduced in Geoffrey Moore's *Crossing the Chasm*, refers to the critical transition between early adopters and the mainstream market, where startups must overcome niche adoption challenges to achieve widespread success and rapid growth.
Churn Rate	The churn rate measures the percentage of customers who stop doing business with a company over a given period, with high churn signaling low retention and weak product-market fit, making it crucial for early-stage startups to address before focusing on growth.
Clawback	A clawback provision is a contractual clause, commonly used in employment agreements within financial firms, requiring the repayment of previously disbursed funds if specific conditions are met.
Cliff	A cliff is the waiting period before employees or founders can access their vested shares, promoting long-term commitment. In a 4-year vesting plan with a 1-year cliff, leaving within the first year forfeits any equity.
Closed-Loop	Closed-loop control uses real-time feedback from sensors to continuously adjust a robot's actions, ensuring precise movement and task accuracy by monitoring factors like position, speed, and force.
Collaborative Robot	A collaborative robot, or cobot, is designed to work alongside humans, operating in one or more of four specific collaborative modes to ensure safe and efficient interaction.
Collective Learning	Collective learning is an AI training strategy that draws on the shared knowledge and expertise of multiple models, leading to more sophisticated and robust intelligence.

Come Along Rights	Also known as Tag-Along Rights, this provision allows investors to sell their shares if a founder or key employee sells theirs, ensuring fair liquidity and protecting investors from being left behind after key stakeholder exits.
Common Stock	Common stock, typically granted to founders, management, and employees, is an equity security subordinate to preferred shares in liquidation events.
Contact Sensor	A device that detects objects or measures force and torque through physical contact, used to determine an object's location, identity, and orientation.
Control Device	Any piece of control hardware that provides a means for human intervention in the control of a robot or robot system, such as an emergency-stop button, a start button, or a selector switch.
Control Rights	Control rights grant investors or shareholders authority over company decisions, typically involving board seat appointments, voting power, and requiring consent for key actions like incurring debt.
Controllability	Controllability is the ability to regulate and guide an AI system's decision-making, ensuring accuracy, safety, ethical behavior, and minimizing unintended consequences.
Controller	The robot control mechanism, typically a computer, manages data, programs, algorithms, and logic to govern operations, ensuring precise and effective functionality.
Conversational AI	A subfield of AI that enables systems to understand, generate, and engage in natural, human-like conversations.
Conversion Rate or Ratio	The number of common stock shares each preferred share can be converted into.
Convertible Debt	Convertible debt notes allow investors to exchange their loan and interest payments (depending on the terms) for equity in early-stage at the next qualified financing, often at a discount.
Covenant	A covenant is a binding legal obligation in a contract to perform a specific action, while a Negative Covenant requires refraining from certain actions.
Crowdfunding	Crowdfunding leverages small contributions from a large group to fund projects or businesses, including pre-sales (e.g., Indiegogo) or equity-based funding through platforms like Republic and OurCrowd.
Cybernetic	In robotics, cybernetics focuses on control and feedback systems that enable robots to sense, adapt, and act autonomously, mimicking the adaptive behavior of living organisms.

DARPA	The Defense Advanced Research Projects Agency (DARPA), founded in 1958 as ARPA in response to the Soviet launch of Sputnik, has driven U.S. innovation in robotics, AI, and defense, creating technologies like spy drones and humanoid robots to maintain national security superiority.
Deal Lead	The deal lead is the investor who defines terms, manages due diligence, oversees the investment process, and typically takes a board seat.
Deep Learning	A cutting-edge subfield of machine learning that leverages multi-layered neural networks to learn from vast data sets.
Degrees of Freedom	A degree of freedom (DOF) in robotics refers to each independently movable component, typically powered by one actuator, with more degrees of freedom enhancing versatility but increasing control complexity.
Deterministic Model	A deterministic model adheres to a precise set of rules and conditions, ensuring a clear cause-and-effect process that leads to a singular, predictable outcome.
Dilution	As outside investments grow, founders' equity diminishes or dilutes, but remember owning 10% of a billion-dollar company far outweighs owning 100% of nothing (a natural tradeoff in building a startup).
Dogfooding	Dogfooding is when a company has its employees use its own products before release. This internal testing helps discover bugs and make improvements.
Down Round	A down round occurs when a startup raises capital by selling shares at a lower price per share than in a previous funding round. T
Downtime	Downtime refers to the period when a robot or production line is non-operational, typically due to malfunction or system failure, leading to a temporary halt in production
Drag-Along Rights	A contractual provision granting majority shareholders the authority to compel minority shareholders to participate in the sale, merger, or transfer of the company under the same terms and conditions, ensuring uniformity in the transaction.
Drone	Drones, or Uncrewed Aerial Vehicles (UAVs), originated in the 1930s with the British Royal Navy's "Queen Bee" for military training and have since evolved into essential tools for military, commercial, and recreational applications.
Dynamics	Dynamics studies motion and the forces that cause it, and in robotics, it examines the complex interactions between a robot arm's moving parts, forces, and kinematics.

EBIT	EBIT (Earnings Before Interest and Taxes) measures a company's profitability from operations, while EBITDA (Earnings Before Interest, Taxes, Depreciation, and Amortization) excludes non-cash expenses to focus on operational performance; companies are often valued based on multiples of EBIT or EBITDA to reflect their earning potential.
Elevator Pitch	An elevator pitch is a brief, compelling presentation designed to quickly outline an idea or opportunity and capture interest within the span of an elevator ride.
Emergency Stop	An emergency stop (e-stop) is a hardware feature that quickly halts equipment by cutting power for safety.
Employee Stock Ownership Program	An employee stock ownership program (ESOP) is a pool of stock options that grants employees equity in the company, aligning their interests with its success and offering financial rewards as its value grows.
Encoder	An encoder is a feedback device that converts a robot's motor shaft position into digital signals, using coded disks to track movement for precise control.
End-Effector	An end-effector is the tool attached to a robot's arm, allowing it to perform tasks such as gripping, welding, or painting.
Exercise Price	The exercise price, or strike price, is the fixed amount an option holder pays to buy shares when exercising stock options, set at the shares' Fair Market Value (FMV) at grant and unchanged through the option's life.
Exit	An exit is when a founder or investor leaves a business through a liquidity event, such as an acquisition or IPO, allowing shareholders to sell their shares.
Exoskeleton	applications in medicine for rehabilitation, in industry to prevent fatigue and injury, and in the military to boost strength and endurance.
Explainability	Explainability uses techniques to make AI decision-making transparent and understandable, fostering trust, accountability, and informed use of AI systems.
Extensibility	Extensibility in AI refers to the system's ability to adapt and expand its capabilities to new domains, tasks, or datasets with minimal retraining or architectural changes.
Extraction	Extraction is a generative model's ability to analyze large datasets, identify relevant patterns, trends, and specific information, and distill meaningful insights for accurate, context-aware outputs.
Fair Market Value	The price a willing buyer would pay for a company or asset, typically determined by recent transactions involving comparable companies or assets, particularly for private or illiquid holdings.

Feedback Control	Feedback control is a system where sensor data on a robot's position, movement, or environment is continuously sent to the controller, enabling real-time adjustments for precise, adaptive, and accurate performance.
Fiduciary	In the startup context, a fiduciary is someone, such as a founder, board member, or investor, entrusted to act in the best interests of the company and its shareholders.
First Mover Advantage	First mover advantage refers to a startup's ability to gain a competitive edge by being the first to market in a new product category.
First Refusal Rights	First refusal rights, or preemptive rights, allow early investors to purchase shares in future funding rounds or founder shares before external sales, helping them maintain ownership and influence as the startup grows.
Flat Round	A flat round occurs when a startup raises funding at the same valuation as the previous round, signaling limited growth and raising concerns about scalability, market traction, or future potential, which can deter new investors.
Flexibility	In robotics, flexibility is a robot's ability to adapt to various tasks without extensive reprogramming, such as a robot arm seamlessly switching between assembly, material handling, and quality inspections, enhancing efficiency and reducing downtime.
Follow-On Funding	Follow-on funding occurs when VC firms invest in later rounds of startups they've already backed, maintaining equity stakes, signaling confidence, and supporting growth and scaling efforts.
Force Feedback	Force feedback is a sensing technique where end-effector sensors relay real-time data to the control unit, enabling precise force adjustments for delicate tasks in robotics and tactile sensations in haptic systems.
Foundation Models	Foundation models, like large language models (LLMs), are powerful AI systems trained on vast data, driving versatile applications across domains but demanding high development and deployment costs.
Founders Shares	Founder shares grant originators ownership and control, often with enhanced voting rights, while attracting talent through equity alignment and offering potential long-term tax benefits.
Freemium	A business model offering a free basic product, with paid upgrades for premium features.
Friends And Family Round	The friends and family round, a startup's first funding from close connections, fuels early ideas or MVPs with flexible terms but demands careful expectation management.

Full Ratchet	Full ratchet is an anti-dilution mechanism that protects early investors by adjusting their shares to match lower future prices or granting additional shares, preserving their ownership but significantly diluting founders' stakes.
Fully Diluted	Fully diluted refers to the total number of issued common shares, including those issuable from options, warrants, convertible preferred stock, and debt, used to compare ownership percentages across different security classes by converting them to common stock equivalents.
Gantry Robot	A gantry is an adjustable machine that moves precisely along X, Y, and Z axes on fixed tracks, offering cost-effective solutions for large areas, heavy loads, and automation in manufacturing and logistics.
General Partner	A general partner (GP) is a senior manager at a venture capital fund who leads investment decisions and strategy, often holding a key vote in the investment committee.
General Solicitation	General solicitation is the public advertising of investment opportunities, which can attract more investors but may forfeit certain legal protections, making it vital for startup founders to balance regulatory compliance with broadening their investor reach.
Generative AI	Generative AI is revolutionizing robotics, enabling robots to learn from vast data, adapt to dynamic environments, and enhance decision-making, creativity, and problem-solving, unlocking real-time collaboration and advanced automation across industries.
GPS	The Global Positioning System (GPS) determines precise coordinates using satellite signals but struggles in indoor or GPS-denied environments, prompting robots to enhance accuracy by integrating technologies like inertial navigation or local positioning systems.
GPU	A GPU (graphics processing unit) is a specialized processor optimized for parallel computing, enabling real-time processing of complex data to accelerate deep learning tasks like image recognition, object detection, and decision-making, driving AI-powered robotics automation.
Grandfather Rights	Grandfather rights, or clauses, let existing agreements persist under old rules despite new regulations, offering startup founders stability, competitive advantage, and strategic flexibility during regulatory changes.
Gripper	A gripper, an end effector on a robotic arm, uses mechanisms like fingers, magnets, or suction for precise manipulation, with types suited to diverse industrial tasks.

Gross Margin	Gross margin, the percentage difference between revenue and cost of goods sold (COGS), reflects profitability and efficiency, with VCs favoring SaaS companies for their high margins (70%+) and scalability, unlike robotics firms with lower margins (30-50%) due to higher manufacturing costs.
Growth Equity	Growth Equity refers to private investments in late-stage companies aimed at financing revenue growth through market expansion.
Growth Hacking	Growth hacking involves using unconventional and cost-effective digital marketing strategies to quickly grow and retain a user base, boost brand exposure, and drive product sales.
Gyro	A gyro, or gyroscope, uses rotating, vibrating, or optical mechanisms to detect orientation changes, enabling robots to maintain stability and precise control by providing real-time angular velocity data for efficient navigation.
Hallucination	Hallucination in AI occurs when a system, particularly in natural language processing, generates irrelevant or incorrect outputs due to insufficient context, over-reliance on training data, or misinterpretation, underscoring the challenges of accurate language comprehension.
Haptics	Haptics studies touch to improve control systems, using devices that apply forces and vibrations to convey robot actions or translate operator inputs, such as force feedback in joysticks or haptic gloves controlling robot hands.
Harmonic Drive	A harmonic drive is a lightweight, compact speed reducer that converts high-speed, low-torque input into low-speed, high-torque output, using an elliptical cam and flexible spline for precise, backlash-free motion, commonly used on smaller robotic joint axes for high-accuracy tasks.
Harness	A harness bundles wires to deliver power and signals between devices, connecting motors, sensors, and controllers in robotics to ensure efficient communication, simplify maintenance, and enhance system reliability.
Hockey Stick	The term "hockey stick" describes a startup's growth curve—modest at first, then exponential—signaling product-market fit, scalability, and profitability, driven by strong traction, validated business models, and effective market timing, making it attractive to investors.
Home Position	Home Position is a fixed reference point along a manipulator's coordinate axes, marked for calibration and essential for accurate movement, programming, repeatability, and operational efficiency in robotic applications.

Humanoid	A humanoid robot, designed with human-like features such as arms and a torso, prioritizes functionality over aesthetics to perform tasks requiring human-like interaction, making it valuable in sectors like research, healthcare, logistics, and manufacturing for improving efficiency and productivity.
Hydraulic	Hydraulic actuators, powered by pressurized fluid, provide exceptional strength for heavy lifting and forceful tasks in applications like construction, industrial robotics, automotive systems, aerospace, and marine operations, despite slower speeds and maintenance challenges.
ICRA	ICRA, the IEEE International Conference on Robotics and Automation, is a leading global robotics conference held annually, where researchers and professionals share advancements in robotics and automation, with its name pronounced "e-cra" as part of its identity.
IEC	IEC, the International Electrotechnical Commission, is a global body that develops standards for electrical and electronic technologies, ensuring safety, efficiency, interoperability, and reliability across industries.
IMU	An IMU, or Inertial Measurement Unit, combines accelerometers, gyroscopes, and compasses to provide real-time data on orientation, velocity, and motion, enabling precise navigation and movement in UAVs, mobile robots, and humanoids.
In-Kind Support	In-kind support refers to non-monetary resources, including mentorship, strategic guidance, industry connections, sales support, investor introductions, and services like cloud hosting credits or pro bono consulting, often provided by VC platform teams to enhance startups' capabilities, foster collaboration, and accelerate growth.
Information Rights	IInformation rights grant investors regular financial updates and access to corporate records, such as stock ledgers and books, under Delaware law and similar regulations, provided requests are made for a "proper purpose," ensuring transparency, accountability, and informed decision-making.
Input Devices	Input devices like joysticks, keyboards, and panels enable intuitive human-machine interaction, improving control and efficiency in robotics.
Inside Round	An inside round is a financing round exclusively involving existing investors, reflecting their confidence in the company's growth, streamlining negotiations, minimizing dilution, and showcasing strong support and alignment with the founders.

Instruction	An instruction is a line of code directing a system controller to perform actions—such as moving, processing data, or executing algorithms—enabling robots and AI to interact with their environment and perform tasks from simple movements to complex problem-solving.
Integrator	An integrator, or systems integrator, is a company that provides value-added services to create comprehensive automation solutions by combining robots with other automation and control equipment.
Intelligence Amplification	Intelligence amplification enhances human capabilities by combining AI systems with traditional tools to boost decision-making, problem-solving, and productivity.
Inverse Kinematics	Inverse kinematics is a mathematical framework that calculates joint movements for precise robotic positioning, critical for tasks like manipulation and navigation in applications such as robotic arms and autonomous vehicles.
Investment Syndicate	An investment syndicate is a group of investors pooling resources to fund a company, sharing risks and expertise while sometimes offering startups additional support, credibility, and mentorship, though reliance on syndicates alone may deter venture capital interest.
Investor's Rights Agreement	An Investor's Rights Agreement is a legal document protecting investors' interests by granting rights such as participation in future funding rounds, board meeting observation, and first refusal on shares, fostering transparency and safeguarding their investment.
IROS	IROS, or the IEEE International Conference on Intelligent Robots and Systems, is one of the world's largest robotics conferences.
ISO	ISO, or the International Organization for Standardization, is an independent, non-governmental international organization that develops and publishes globally recognized standards across various industries.
Issued Shares	Issued shares represent the total number of authorized shares, including both common and preferred stock, distributed to stockholders, reflecting ownership in the company and its equity structure, distinct from authorized shares, which are the maximum allowed by the corporate charter.
Issuer	Issued shares represent the total number of authorized shares, including both common and preferred stock, distributed to stockholders, reflecting ownership in the company and its equity structure, distinct from authorized shares, which are the maximum allowed by the corporate charter.

Jacobian Matrix	The Jacobian matrix links the rates of change of a robot's joint variables to its end-effector coordinates, playing a key role in controlling positioning, motion, and tasks like trajectory planning, inverse kinematics, and real-time control.
Joint	A joint is a key component of a manipulator system that enables rotation and movement, providing flexibility and degrees of freedom essential for the robot's functionality and dexterity in performing various tasks.
K1	Schedule K-1 is an IRS tax form issued annually to report an investor's share of earnings, losses, deductions, and credits from partnerships, crucial for VCs and startups to understand for tax liabilities, financial reporting, and compliance.
Key Man Clause	The key man clause allows a startup to terminate an agreement if a key individual, such as a founder or essential team member, leaves, protecting investors from potential leadership changes that could impact the company's success.
Kinematics	Kinematics studies the relationship between a robot's end effector and its joints, using linear functions for Cartesian robots and complex trigonometric calculations for revolute robots, and is divided into forward and inverse kinematics to ensure precise control and programming.
Knowledge Generation	Knowledge generation uses machine learning and algorithms to analyze data, identify patterns, and produce insights, driving innovation and improving decision-making and predictive capabilities.
Laser	In the realm of robotics, lasers serve as sophisticated non-contact sensors, enabling a myriad of applications such as distance measurement, precise location identification, surface mapping, barcode scanning, cutting, and welding.
Latency	Latency is the time delay between input and output in a system, and minimizing it is crucial for ensuring responsive AI applications and smooth user experiences, especially in real-time scenarios.
Lean Startup	TThe Lean Startup methodology encourages efficient product development through validated learning, rapid experimentation, and iterative design, emphasizing minimum viable products (MVPs) to test assumptions, gather feedback, and reduce financial risk, making it particularly beneficial for hardware startups to improve product-market fit and minimize costly failures.
Letter of Intent	A Letter of Intent (LOI) is a non-binding agreement outlining terms, pricing, and commitments between startups and potential customers, building investor confidence by showing secured interest, and often used alongside Memoranda of Understanding (MOUs) in larger projects.

LIDAR	LIDAR (Light Detection and Ranging) is crucial for robotics and autonomous vehicles, using lasers to create high-resolution 2D or 3D maps, detect obstacles, and enable real-time decision-making, while enhancing perception when combined with other sensors like cameras and radar.
Liquidation	Liquidation is the process of converting a company's assets into cash, where payouts are governed by preference clauses, prioritizing debt holders and preferred shareholders over common stockholders.
Liquidation Preference	Liquidation preference determines the order in which investors and debt holders are compensated during a company's liquidation or bankruptcy, protecting venture capitalists' investments and clarifying potential recovery.
Liquidation Waterfall	The liquidation waterfall defines the payout sequence during a company's liquidation, prioritizing debt holders and preferred shareholders over common shareholders to clarify asset distribution and potential returns.
Liquidity	Liquidity refers to the ease with which an asset can be bought or sold in the market without significantly affecting its price.
Liquidity Event	Liquidity events are critical for startups and their investors as they provide a tangible ROI and signify significant growth milestones, such as acquisitions or initial public offerings (IPOs).
Load Cycle Time	Load Cycle Time measures the total time to unload the last workpiece and load the next, encompassing all transition activities and serving as a key metric for improving operational efficiency and productivity.
Lock-Up Period	A lock-up period is a specified timeframe during which the holder of a security, such as shares from an initial public offering (IPO), cannot sell or transfer that security.
Low-Code	Low-code is a visual development approach that enables users to create applications with minimal hand-coding through intuitive graphical interfaces and pre-built components.
LTV	Lifetime Value (LTV) is a crucial metric that represents the total revenue a startup can anticipate earning from a customer throughout their relationship with the company (**LTV =** Average Purchase Value **x** Average Purchase Frequency **x** Customer Lifespan).
Machine Learning	Machine Learning (ML) enables machines to improve through experience and data analysis, driving robotics applications like autonomous navigation, predictive maintenance, and real-time object recognition, enhancing their adaptability, efficiency, and effectiveness.

Magnetic Detectors	Magnetic detectors are sensors that detect ferromagnetic materials with high precision, used in robotics for applications like metal detection on assembly lines, obstacle avoidance, and treasure hunting.
Major Investor	A major investor is an individual or entity that holds a significant ownership stake in a company, granting them enhanced rights and influence within the organization.
Manipulator	A manipulator is a robotic mechanism composed of interconnected segments that move relative to one another, enabling the graceful grasping and manipulation of objects with multiple degrees of freedom.
Market Penetration	Market penetration is a metric that measures the percentage of a product or service utilized by customers relative to the total estimated market size for that offering.
Material Handling	By automating material handling tasks, robotic arms and mobile robots enhance productivity, reduce labor costs, and improve safety by minimizing human intervention in potentially hazardous environments.
Mechatronics	Mechatronics is an interdisciplinary field that integrates mechanical engineering, electrical engineering, and computer science to design and develop advanced robotics and intelligent machines.
Merger	A merger is the process of combining two distinct companies to form a single new entity.
MFN	Most Favored Nation (MFN) is a provision that ensures a lead investor—typically the largest backer of a startup—receives all the benefits outlined in any agreements made with other investors.
Milestone	A milestone is a significant achievement that triggers additional investments from venture capitalists, and for startups, clearly defined milestones like product launches, user acquisition targets, or revenue benchmarks enhance credibility and attract further funding by demonstrating progress and growth potential.
Moat	A moat is a metaphor for a company's capacity to maintain a competitive advantage over its rivals, safeguarding its market share and profitability.
Model Chaining	Model chaining is a data science technique that links multiple machine learning models sequentially to enhance predictions by building on each model's output, improving accuracy and capturing complex relationships in data.

Modularity	Modularity refers to the design principle that enhances the flexibility of a robot and its control system by using separate, interchangeable units.
Monthly Recurring Revenue	Monthly Recurring Revenue (MRR) is a key metric for subscription-based businesses, calculated by multiplying active subscriptions by the average revenue per user (ARPU), and is crucial for robotics-as-a-service (RaaS) companies, offering steady revenue and insights into growth potential, scalability, and investment viability.
Motion Planning	Motion planning is a technique that breaks down desired movements into discrete steps, calculating optimal paths for mobile robots to navigate obstacles or determining precise joint movements for manipulators, thereby enhancing accuracy and effectiveness in various robotic applications.
Multi-Hop Reasoning	Multi-hop in natural language processing refers to an AI model's ability to synthesize answers by connecting information from multiple sources, enhancing its reasoning and comprehension to address complex questions with greater accuracy.
Multimodal Language Model	Multimodal Language Models are deep learning models that process and understand diverse data types, including text, images, audio, and video, enabling them to generate richer, contextually aware outputs and perform complex tasks like image captioning and video analysis.
Muting	Muting is the intentional deactivation of safety devices during robot testing to allow full operation for performance assessment, but it must be done cautiously with proper safeguards to ensure a safe environment and protect personnel.
Minimum Viable Product	A Minimum Viable Product (MVP) is a basic version of a product with essential features designed to attract early adopters, validate a business idea, and gather user feedback, helping startups minimize risk, refine the product, and pivot based on market response.
Natural Language Processing	Natural Language Processing (NLP) is a subfield of AI that enables computers to understand and process free-flowing language, supporting tasks like sentiment analysis, translation, chatbots, and information retrieval to enhance human-computer interactions and drive innovation.
Net Revenue	Net revenue refers to the total income generated by an AI or robotics startup after accounting for deductions such as discounts, price reductions, and refunds.

Neural Network	A neural network is a machine learning model inspired by the human brain, composed of interconnected nodes that excel at processing complex data patterns, enabling advancements in autonomous vehicles, drones, robots, and generative AI by enhancing tasks like object recognition, image analysis, motion planning, and media creation.
No-Code	No-code is an approach to designing and deploying applications that eliminates the need for coding or programming knowledge, allowing users to create software solutions through intuitive visual interfaces and drag-and-drop features.
Non-Binding	Non-binding refers to documents that do not create enforceable obligations, such as term sheets, MOUs, and LOIs, allowing either party to withdraw without legal consequences, offering flexibility during negotiations, with investors and startups typically formalizing agreements after mutual consent.
Non-Disclosure Agreement	A non-disclosure agreement (NDA) is a legally binding contract designed to protect the confidentiality of sensitive information shared by entrepreneurs with third parties, such as investors and strategic partners.
Non-Equity Assistance	Non-equity assistance is a funding model where entrepreneurs raise capital without giving up ownership, typically through a repayment structure that compensates investors with a multiple of their initial investment, allowing startups to retain control while minimizing dilution.
Opex	Operating Expense (Opex) refers to ongoing costs for daily operations, and understanding it is key when selling robotic systems, as these solutions can reduce labor and maintenance costs, offering long-term savings and enhanced efficiency, which can create compelling value propositions for potential customers in the automation industry.
Optical Proximity Sensors	Optical proximity sensors measure distance by detecting reflected light, often using lasers for high precision, and are essential in robotics for navigation, obstacle avoidance, and accurate object interaction across applications like industrial automation
Options	Option grants for startup employees provide the right to purchase shares at a set price after a vesting period, typically including a one-year cliff, with gradual vesting thereafter, incentivizing long-term commitment and aligning employee interests with the company's growth.

Option Pool	An option pool is a reserved set of stock options created by a startup to attract and retain talent, typically representing 10% to 20% of the company's total shares.
Palletizing	Palletizing and de-palletizing, the processes of stacking and unloading packages onto pallets, are increasingly automated in e-commerce with robotic systems that enhance speed, accuracy, and efficiency, streamline warehouse operations, reduce labor costs, and improve supply chain performance.
Parallel Robot	A parallel robot, such as the Delta robot with three arms connected to a triangular platform, is designed for high-speed, precise operations, excelling in industries like automotive and food processing for tasks such as assembly and packaging, with its versatility and speed making it essential in modern manufacturing and logistics.
Pari Passu	Pari passu is a legal term meaning "on the same terms as," ensuring equal treatment for all parties in an agreement, which is crucial for startups during fundraising to promote fairness in rights, obligations, and equity ownership, fostering transparency and trust with investors.
Participating Preferred	Participating preferred stock allows holders to receive their liquidation preference first and then share in remaining assets, offering "double dipping" benefits, making it an attractive option for startups to secure investors and provide upside potential.
Party Round	A party round is a funding round where a startup secures capital from multiple investors with minimal stakes, often facilitated by Special Purpose Vehicles (SPVs) to aggregate smaller investments and attract diverse backers, in contrast to traditional rounds with a lead investor.
Pay to Play	In venture capital, a "pay-to-play" provision requires investors to maintain their pro rata commitment in future funding rounds, or risk losing key rights, such as anti-dilution protections, ensuring ongoing support, stability, and enhancing the startup's valuation and appeal.
Payload - Maximum	The maximum mass that the robot can manipulate or otherwise move at a specified speed, acceleration/deceleration, center of gravity location (offset), and repeatability under continuous operation over a specified working space.
Pick and Place	Pick and Place refers to a repetitive task in which a manipulator picks up an object and places it in a desired location before returning to its rest position.

Piggyback Registration Rights	Piggyback Registration Rights allow investors to include their shares in a company's public offering, whether for themselves or another shareholder, typically applying to unlimited offerings until terminated, and enabling participation in share registrations for an IPO.
Pivot	Pivot refers to a strategic shift in a business model or approach, driven by changes in industry dynamics, customer preferences, or new market insights. .
Play Mode	Play Mode is the operational state of a robot after programming in Teach Mode, where the robot executes programmed tasks in real-time, performing designated functions to enhance efficiency and productivity in industrial environments.
Pledge	A pledge is an investor's commitment to provide capital to a startup or VC fund, promising funds when needed, typically during fundraising or in stages.
Pneumatic Actuator	A pneumatic actuator uses compressed air or other gases to create mechanical motion, harnessing pressurized gas for precise control and rapid response in applications ranging from industrial machinery to automation systems.
Point-to-Point	Point-to-Point is a robotic motion mode where movement is defined by specific points along a path, enabling direct transitions between points, and is particularly valuable in applications like assembly, pick-and-place, and material handling, where precision, repeatability, and efficiency are essential.
Portfolio Company	A portfolio company is a business that has received investment from a venture capital fund, holding company, or startup studio, gaining not only financial support but also strategic guidance, resources, and access to networks that enhance growth, innovation, and success.
Post-Money Valuation	The post-money valuation of a company is calculated by adding the investment amount to the pre-money valuation, reflecting the market value after the investment, such as a $3 million investment in a company with a $10 million pre-money valuation resulting in a $13 million post-money valuation.
Power and Force Limiting	Power and Force Limiting (PFL) is a collaborative feature that ensures safe interaction between operators and robots by slowing down and stopping the robot before contact, requiring functional safety measures and detection hardware for effective implementation.
Private Placement Memo	A Private Placement Memo (PPM) provides detailed investment information, a pitch deck offers a concise startup overview, and a business plan outlines strategy and financial forecasts, each playing a key role in investment and business development.

Preemptive Rights	Preemptive Rights grant existing shareholders the opportunity to purchase new shares before they are offered to others, protecting their ownership from dilution and ensuring they can invest in future financing rounds on favorable terms, often outlined in term sheets as "Right to Participate Pro Rata in Future Rounds."
Preferred Stock	Preferred Stock is a class of equity security that grants shareholders specific rights over common stockholders, including liquidation preferences, which ensure that investors receive distributions of money or assets before common stockholders in the event of a sale, merger, or liquidation.
Presence-Sensing Safeguarding Device	resence-sensing devices create a safety field to detect intrusions by people, robots, or objects, enhancing robotic safety by providing real-time alerts or responses in environments like collaborative workspaces, automated warehouses, robotic surgery, and security systems.
Price Anti-dilution Protection	Price anti-dilution protection safeguards investors by adjusting the conversion ratio of preferred shares if the company issues stock at a lower price, with full ratchet and weighted average (broad-based or narrow-based) methods offering varying levels of protection to maintain investment value despite valuation decreases.
Priced Round	A priced round is a funding round where investors purchase newly issued shares at a set price based on the company's pre-investment valuation, providing clarity on ownership stakes and financial terms, and is typically preferred in later-stage funding to establish clear expectations and boost investor confidence.
Private Equity	Private equity refers to investment funds, typically structured as limited partnerships, that acquire and restructure companies to improve their performance and increase their value.
Pro-Rata Rights	Pro-rata rights grant investors the ability to participate in subsequent funding rounds, enabling them to maintain their percentage ownership in the company. This mechanism allows investors to continue supporting companies they believe in, ensuring they can increase or preserve their investment stake as new capital is raised.
Probabilistic Model	A probabilistic model uses probabilities to inform decision-making, incorporating uncertainty and variability to enable more nuanced predictions and adaptive responses in complex environments.
Proof-of-Concept	In startup funding, a proof-of-concept (POC) validates a product's viability and market fit, turning interest into orders and driving growth.

Protective Provisions	Protective provisions grant investors the right to veto certain company actions, ensuring their interests are safeguarded by requiring approval through a class vote of Preferred Stock for significant decisions that could impact the company's value or direction.
Prototype	A prototype is an initial model of a robotic or AI system developed to test and validate its design and functionality.
Proximity Sensor	A proximity sensor detects nearby objects without contact using technologies like radio frequency, ultrasonic, and photoelectric methods, and is crucial in automation, robotics, and industrial processes for applications such as counting, metal detection, and level control, enhancing efficiency and safety.
Quadrotor	A quadrotor, or "drone," is a flying robot with four horizontal rotors for lift and control, known for its simplicity, affordability, and versatility in applications ranging from recreational use to professional filming, surveying, and search-and-rescue, with larger models featuring more rotors for enhanced performance.
Quality Assurance	Quality assurance (QA) encompasses the methods, policies, and procedures necessary for conducting thorough quality testing during the design, manufacturing, and delivery phases of creating, reprogramming, or maintaining robots.
Radar	Radar is a sensor technology that uses radio waves to measure the distance and location of objects, excelling in various weather conditions and long-range detection, making it crucial for applications like autonomous vehicles, drones, aviation, and military surveillance for obstacle detection, navigation, and collision avoidance.
Ramen Profitable	A startup is considered "ramen profitable" when it achieves minimal profitability, generating just enough revenue to cover the basic living expenses of its founders, often akin to subsisting on a frugal diet of ramen noodles.
Ratchet	A ratchet provision is a contractual safeguard that adjusts an investor's ownership stake in the event of a down round or underperforming IPO, ensuring their shares convert to common stock at a predetermined value, preserving their rights and mitigating the impact of declining company valuations.
Reach	Reach refers to the operational area a robot's end-effector can access in at least one orientation, crucial for optimizing design and ensuring efficient interaction with objects and task performance within its intended environment.
Real-Time System	A real-time system is a computer system designed to execute tasks within specified time constraints while simultaneously supporting associated processes.

Reasoning	AI reasoning enables systems to analyze data, recognize patterns, and make informed decisions, enhancing adaptability and problem-solving across diverse tasks and domains.
Recapitalization	Recapitalization, particularly in the context of down rounds, can be seen as a negative indicator, as it often reflects a declining valuation, loss of investor confidence, increased dilution for existing shareholders, financial instability, and potential operational challenges, raising concerns about the startup's long-term viability and management effectiveness.
Redemption Rights	Redemption rights give investors the option to require a startup to repurchase their shares at a specified price after a set period, providing leverage for liquidation if no IPO or liquidity event occurs, while boosting investor confidence and attracting capital by offering an exit strategy.
Reinforcement Learning	Reinforcement learning is a type of machine learning where an autonomous agent learns to make decisions by interacting with its environment and receiving feedback, enabling robots and self-driving vehicles to adapt and optimize performance based on real-time experience.
Reliability	Reliability refers to the probability that a device, such as a robot, operates without failure over a specified period, measured by uptime or Mean Time Between Failure (MTBF), and is crucial for assessing performance, durability, and minimizing downtime in applications requiring continuous operation.
Representations and Warranties	Representations and warranties are formal statements in a securities purchase agreement where a startup affirms its operations, financial condition, and legal compliance, providing transparency and mitigating risks for investors while fostering trust and accountability.
Repurchase Option	A repurchase option allows a company to buy back vested or issued shares from shareholders at a predetermined price or under specific conditions, enabling strategic management of shareholder composition, financial flexibility, and market responsiveness.
Resolution	Resolution refers to the minimum motion required at a robot joint for the position sensing system to register a change, with variability in the end effector's resolution in world coordinates, especially in revolute arms, affecting the precision and performance of the robotic system depending on orientation and arm configuration.
Responsible AI	Responsible AI focuses on developing and using AI systems ethically, with transparency and accountability, to ensure positive outcomes for employees, businesses, customers, and society, fostering trust and enabling scalable, impactful innovations aligned with societal values.

Restricted Stock	Restricted stock, or Restricted Stock Units (RSUs), is a class of equity with transfer or sale restrictions, often used as employee compensation to incentivize retention and performance, with terms varying by company and shareholder.
Return on Investment	ROI measures the value received by customers relative to costs and the financial returns earned by investors on capital, with high ROI indicating successful performance, value creation, and fostering loyalty while attracting further investment.
Revenue	Revenue is the total amount of money a company generates from its core business activities during a specific period, before any expenses are deducted.
Revenue Multiple	The Revenue Multiple, often expressed as EV/TTM Revenue (Total Enterprise Value to Trailing Twelve Months Revenue), is a valuation metric commonly used to assess companies that are not yet profitable.
Revenue Run Rate	The Revenue Run Rate annualizes current revenue to predict future performance and growth, providing startups and investors with a forward-looking gauge of financial potential.
Revenue-Based Financing	Revenue-based financing is a funding model where an investor provides capital to a startup in exchange for a percentage of its total revenue or sales over a specified period, rather than equity.
Reverse Dilution	Reverse dilution occurs when unvested shares from former employees are returned to the company, reducing outstanding shares and potentially increasing the ownership percentage of remaining shareholders, which can improve the company's capital structure.
Revolute Joint	Rotary joints are the articulating components in robots that enable rotational movement, providing the flexibility and range of motion necessary for precise positioning, manipulation, and complex tasks in industries like manufacturing, healthcare, and logistics.
Right of First Refusal	The Right of First Refusal (ROFR) allows investors to purchase stock on the same terms offered to a third party before shares can be sold, helping companies retain control over ownership changes and stabilizing the ownership structure.
Roboethics	Roboethics is a field that explores the ethical implications of robotics and AI, focusing on how robots should behave, the responsibilities of their creators, and ensuring that technological advancements align with societal values while minimizing harm and addressing potential societal impacts.

Robot	According to Rodney Brooks' definition in his 2002 book *Flesh and Machines: How Robots Will Change Us*, a robot is "a machine which senses the world, computes, and then decides on some action in the world which has a physical reach beyond itself."
Roll Up	A roll-up is a strategy where investors, often private equity firms, acquire and merge multiple small companies in the same market to achieve cost reductions, operational efficiencies, and increased profitability through economies of scale.
ROS	Robot Operating System (ROS) is an open-source suite of tools, libraries, and middleware that enables robots to perform tasks like vision, navigation, and manipulation, offering flexibility and robustness for rapid development and integration of complex robotic systems.
Rotary Joint	A rotary joint allows twisting, swinging, or bending motion around an axis, enabling rotational movement similar to a human elbow, and is essential in robotics for providing flexibility and enabling precise, agile movements in robotic arms and systems.
ROV	Remotely Operated Vehicles (ROVs) are compact, boxy machines equipped with mechanical arms, commonly used for deep-sea tasks such as inspecting oil rigs and exploring shipwrecks.
Runway	Runway refers to the amount of time a startup can operate based on its cash reserves and burn rate, helping founders plan for budgeting, strategic decisions, and fundraising, while allowing investors to assess financial health and when the company will need to raise capital—ideally before cash reserves are nearly exhausted.
SaaS	Software as a Service (SaaS) is a subscription-based model that provides access to applications over the internet, and this concept has expanded into Robot as a Service (RaaS), offering scalable, high-margin access to robotic systems with recurring revenue, low acquisition costs, and continuous innovation, making it appealing to investors for its predictable cash flow and growth potential.
SAFE	A SAFE (Simple Agreement for Future Equity) is a simple, cost-effective tool for early-stage startups to raise capital without incurring debt, allowing investors to buy shares in the future, and is favored for its speed, ease of use, and suitability for smaller funding rounds.tections to investors, leading to discussions about revised versions.

SCARA Robot	A SCARA (Selective Compliant Assembly Robot Arm) is a cylindrical robot with two parallel rotary joints, offering four degrees of freedom, commonly used in industrial automation for tasks like assembly, packaging, and material handling, providing precise, efficient movement in high-speed manufacturing applications.
Secondary Market	The secondary market allows private company stock to be sold to another private party, enabling founders and early investors to unlock value and provide liquidity through employee stock sales, private transactions, specialized platforms, or venture capital firms, enhancing financial flexibility and aligning with the company's growth strategy.
Senior Liquidation Preference	Senior liquidation preference grants certain shareholders, typically investors, higher priority in receiving returns during a company's liquidation or sale, ensuring they recoup their investment before common stockholders, and is crucial for startups to understand as it affects capital structure, shareholder returns, and investment terms in exit scenarios.
Sensory Feedback	Sensory feedback is the data collected by sensors and transmitted to the controller in a closed-loop system, enabling real-time monitoring and adjustments to maintain precision and reliability by correcting errors and adapting to changing conditions.
Separation Agreement	A separation agreement outlines the rights and terms applicable when an employee amicably departs, typically covering severance pay, non-disparagement, non-disclosure, and equity vesting, protecting both the company and employee by clarifying ongoing obligations and rights.
Sequence Modeling	Sequence modeling is a subfield of natural language processing (NLP) that analyzes and predicts sequential data, such as text, speech, or time series, enhancing applications like language translation, speech recognition, and text generation by capturing temporal and contextual relationships.
Series Elastic Actuator	A series elastic actuator combines a motor, gearbox, and spring to introduce compliance, allowing the robot to control movements through force sensing, enhancing adaptability, safety in human interactions, and performance in dynamic environments requiring precise force management.
Servo	A servo is a precision motor with sensors that continuously adjust its movement to ensure accurate, real-time control. Essential in robotics, it powers precise joint movements, grippers, and actuators, delivering high responsiveness and accuracy.

Shareholder Agreement	A shareholder agreement outlines the operation of a company, detailing shareholders' rights and obligations, protections for minority shareholders, and the responsibilities of founders and the board, while ensuring transparency and governance, and complying with regulations like Section 12(g) of the Exchange Act regarding SEC registration.
Shares Outstanding	Shares outstanding refers to the total number of a startup's stock held by all shareholders, including institutional investors and executives, and is crucial for assessing ownership structure, managing equity compensation, and calculating key financial metrics like market capitalization and earnings per share (EPS).
Side Letter	A side letter is an agreement between a startup and an individual investor that specifies additional terms, conditions, or rights not included in the main investment agreement.
Simulation	Simulation refers to a computer program that models a robot and its environment, emulating its behavior to evaluate performance, optimize algorithms, and enhance effectiveness and safety before real-world tasks, reducing costs and refining designs through 3D modeling, kinematics, path planning, and sensor simulations.
Singularity	In robotics, singularity refers to a point where a manipulator becomes uncontrollable due to a loss of degrees of freedom, while also symbolizing a future where AI surpasses human intelligence, raising both technical and philosophical challenges.
SLAM	Simultaneous Localization and Mapping (SLAM) enables robots to create a map of an unknown environment while determining their location, using data from sensors like LIDAR, accelerometers, and IMUs to navigate autonomously in complex, GPS-denied environments, making it essential for autonomous vehicles, robotics, and augmented reality.
Social Proof	Social proof leverages testimonials, reviews, influencer endorsements, and community feedback to build credibility and trust, influencing potential buyers' decisions and enhancing brand authenticity and market perception.
Social Robots	Social robots are designed to engage with humans on a personal level, using advanced communication and emotional intelligence to foster relationships and companionship, making them ideal for home environments, particularly in aging-in-place scenarios, where they can assist with daily tasks, provide companionship, and help seniors maintain independence.

Soft Landing	A soft landing is a controlled entry strategy for startups testing a new market, allowing them to mitigate risk by gradually assessing conditions and adapting their offerings, ultimately paving the way for a more informed and successful expansion.
Sonar	SONAR (Sound Navigation and Ranging) uses ultrasound waves to detect objects and measure distances, offering a more affordable alternative to radar and LIDAR but with lower precision, commonly used in underwater navigation and robotics for obstacle detection.
SPAC	A SPAC is an investment vehicle that raises capital through an IPO to acquire an existing operating company, allowing the target company to go public without the traditional S-1 filing process.
Speech-to-Text	Speech-to-text technology converts spoken language into written text, enhancing human-robot interaction by enabling robots to understand and respond to voice commands, improving efficiency and accessibility in applications like service robots, assistive devices, and autonomous systems.
Stable Diffusion	Stable Diffusion is an advanced artificial intelligence system that employs deep learning techniques to generate high-quality images from textual prompts.
Stacking	Stacking is an AI technique that combines multiple algorithms to improve performance and accuracy by leveraging the strengths of each model, making it particularly effective in applications like image recognition and natural language processing.
Stealth Mode	Stealth mode is a temporary state of secrecy employed by startups to shield their operations from competitors, particularly before a product launch or the initiation of a significant business initiative.
Steerability	AI steerability is the ability to guide an AI system's behavior and outputs according to human intentions and objectives, achieved through techniques like fine-tuning, rule-based systems, and human feedback loops to ensure more reliable, user-aligned collaboration.
Stock Option	A stock option grants the right to buy or sell shares at a predetermined price within a set time, while a stock plan encompasses all rights and interests related to company stock, including bylaws and shareholder agreements, with both serving to align employees' and management's interests with the company's performance, fostering motivation and retention.

Strong AI	Strong AI, or artificial general intelligence (AGI), refers to machines with cognitive abilities similar to human reasoning, capable of learning, understanding, and applying knowledge across diverse domains, enabling complex problem-solving and decision-making that could transform various fields and seamlessly integrate into human life.
Structured Data	Structured data denotes information that is systematically organized and formatted in a standardized manner, making it readily searchable and analyzable by machines.
Success Fee	A success fee is a compensation structure awarded to a broker-dealer upon the successful closure of a transaction, aligning costs with performance and reducing upfront burdens for startups, with only licensed broker-dealers legally eligible to receive the fee.
Super Pro-Rata	Super pro-rata rights allow an investor to acquire more shares than their standard pro-rata allocation in subsequent funding rounds, enabling them to increase their investment and maintain a larger equity stake to capitalize on the company's growth potential.
Supervised Learning	Supervised learning is a type of machine learning where a model is trained on labeled data to identify patterns and make predictions, enabling robots to classify objects, detect obstacles, and understand commands, with applications in autonomous navigation, manipulation, and human-robot interaction.
Swarm Robotics	Swarm robotics, inspired by natural swarms like ant colonies, enables fleets of drones to collaboratively perform tasks such as search and rescue, environmental monitoring, and agricultural surveying, optimizing their movements and adapting to changing conditions for efficient, decentralized operations.
Syndicate	A syndicate, composed of venture capital firms, angel investors, family offices, and strategic partners, pools resources and expertise to share financial risk and increase capital for a startup, similar to a Special Purpose Vehicle (SPV), enhancing due diligence and supporting long-term growth and scalability.
System Integrator	See Integrator.
Tag-Along Right	The right of a minority investor to receive the same benefits as a majority investor. This often applies to a sale of securities by investors and is also known as co-sale right.

Target Market	The Target Market is the specific group of consumers a product or service is aimed at, defined by shared characteristics, and represents a segment of the Serviceable Available Market (SAM) within the Total Addressable Market (TAM), crucial for tailoring marketing strategies and maximizing revenue potential.
Teach	To program a manipulator arm, the operator manually guides it through a sequence of movements, captures each position, and stores the data in the robot controller's memory for later playback, enabling the robot to replicate the motions with precision and efficiency.
Teleoperation	Teleoperation refers to remotely controlling a robot using information from its cameras and sensors, while telepresence extends this by allowing the robot to act as a surrogate for the operator, providing a sense of presence in a remote location and enhancing human interaction with robotic systems through remote control and immersive experiences.
Term Sheet	A term sheet is a nonbinding document outlining the key terms and conditions of a company's fundraising round or investment, summarizing aspects like valuation, investment amount, equity stake, governance, and vesting conditions, and serving as a preliminary agreement to proceed with the transaction, subject to due diligence.
TEV	Total Enterprise Value (TEV) is a comprehensive measure of a startup's value, incorporating market capitalization, total debt, and excluding cash, providing a more accurate representation of worth by accounting for financial obligations and resources, and aiding investors in evaluating potential investments and comparing startups with different capital structures.
Three Laws of Robotics	Formulated by Isaac Asimov in 1942, the Three Laws of Robotics are ethical guidelines to ensure safe human-robot interaction, which are: First Law: A robot may not injure a human being or allow harm through inaction; Second Law: A robot must obey human orders unless it conflicts with the First Law; and Third Law: A robot must protect its own existence, as long as it does not conflict with the First or Second Laws.
Tool	In robotics, a "tool" refers to a working apparatus mounted at the end of a robot arm, such as a gripper or welding torch, and is integral to its functionality, with the Tool & Arm Interference Check Function detecting potential collisions during operation, while the Tool Center Point (TCP), defined as the tip of the tool, serves as the reference for robot motion relative to a fixed coordinate system, facilitating accurate positioning through the tool frame.

Torque

Torque is the rotational force applied to an object, commonly measured in newton-meters (Nm), and in robotics, it describes the power of electric motors, with high torque values being crucial for tasks that require significant force to move heavy objects or overcome resistance.

Touch Sensor

A touch sensor, integrated into a robot's hand or gripper, provides an artificial sense of touch by detecting contact forces with objects, enabling the robot to perceive texture, pressure, and proximity, which is essential for tasks requiring precision and sensitivity.

Trajectory Generation

Trajectory generation calculates motion functions that enable smooth, controlled movement of a robot's joints, ensuring it follows a predefined path while optimizing speed, acceleration, and deceleration for precise, fluid motions and enhanced performance.

Transducer

A transducer is a device that converts environmental signals, such as pressure, temperature, or light, into electrical signals, allowing robots to sense and interact with their surroundings for effective monitoring and automated responses.

Transfer Restrictions

Transfer restrictions are contractual limitations that govern the ability to sell or transfer shares, typically implemented to maintain control over ownership, protect existing shareholders' interests, and ensure regulatory compliance.

Transformer

A transformer is a neural network architecture designed to process sequential data like text, using mechanisms such as self-attention and parallel processing to efficiently capture contextual relationships, and serves as the foundation for models like ChatGPT that excel in natural language processing tasks.

Trough of Disillusionment

The trough of disillusionment, popularized by Gartner's "hype cycles," is a challenging phase for startups marked by setbacks and struggles to achieve product-market fit, where initial excitement fades, exposing gaps between inflated expectations and reality, and often leading to critical reassessments of technologies, recent examples include autonomous vehicles, drones, blockchain, and 3D printing.

TTM

Trailing Twelve Months (TTM) is a financial metric that sums a company's revenue over the past twelve consecutive months, providing a comprehensive view of its performance, smoothing out seasonal fluctuations, and helping investors assess trends, compare peers, and make informed investment decisions.

Turing Test	The Turing Test, proposed by Alan Turing in 1950, evaluates a machine's ability to exhibit human-like responses in conversation, and if a human evaluator cannot reliably distinguish between a human and a machine, the machine is said to have passed, challenging our understanding of intelligence and consciousness, with Eugene Goostman becoming the first AI to claim it passed in 2014.
Two Pizza Rule	The Two Pizza Rule, coined by Jeff Bezos, suggests that no meeting should be so large that two pizzas can't feed the entire group, promoting small, focused meetings that enhance communication, decision-making, and innovation, which is especially important for scaling companies.
UAV	A UAV (uncrewed aerial vehicle) is an aircraft that operates without a human pilot, ranging from remotely controlled systems to fully autonomous drones, and is utilized for a variety of tasks such as surveillance and environmental monitoring.
UI	In robotics, the user interface (UI) refers to the design and layout of the system's control interface, including visual elements like buttons, icons, and menus, that allow users to interact with and command the robot, with a well-designed UI improving ease of use, navigation, and overall user experience in controlling robotic functions.
Uncanny Valley	The uncanny valley is a concept in robotics and AI suggesting that humanoid robots resembling humans too closely but not perfectly can evoke discomfort due to subtle, unsettling differences, as seen with the robot Ava in *Ex Machina* or lifelike dolls and avatars, emphasizing the importance of design for socially acceptable and emotionally engaging robots.
Unicorn	A unicorn is a startup valued at over $1 billion, coined by Aileen Lee to highlight the rarity of such companies, often associated with rapid growth and disruptive innovation, with notable examples like Uber, Airbnb, and SpaceX, signifying both their current success and future growth potential.
Unstructured Data	Unstructured data refers to information that lacks a predefined format, such as text, images, and videos, making it challenging to process and analyze, and includes examples like sensor data from autonomous cars or interactions with humanoid robots, requiring advanced techniques like natural language processing and machine learning to extract valuable insights for decision-making.
Unsupervised Learning	Unsupervised learning allows autonomous robots to analyze unlabeled data, identify patterns, and adapt to dynamic environments, enabling tasks like object recognition and anomaly detection without predefined outcomes.

Uptime	In robotics, uptime refers to the duration a robot is operational and available for tasks, and maximizing uptime is crucial for productivity, efficiency, and continuous operations, as it reflects system reliability and minimizes downtime in applications like manufacturing, logistics, and services.
User Coordinate System	The User Coordinate System (UCS) is a user-defined reference framework anchored to a specific object, enabling efficient programming by defining positions relative to that object, boosting flexibility and productivity in automated processes.
USP	The Unique Selling Proposition (USP) is the distinct benefit a startup offers, setting it apart from competitors by addressing customer needs, attracting potential customers, and driving growth in a competitive market.
UX	User Experience (UX) is the process of designing products that provide meaningful, intuitive, and engaging interactions, focusing on usability, accessibility, and satisfaction to meet user needs and foster loyalty for long-term success.
Vacuum Cup Hand	A vacuum cup hand is a robotic arm end-effector that uses suction to grasp light to moderate-weight objects, such as glass and plastic, providing a secure grip and minimizing slippage in applications like manufacturing and packaging.
Valley of Death	The "valley of death" is a critical phase in a startup's lifecycle, occurring after product launch but before significant revenue, where companies face high operational costs, need ongoing investment, and struggle for market traction, making it a pivotal moment in their journey to profitability and growth.
Value Proposition	The value proposition clearly defines what a startup offers, highlighting its unique benefits and advantages over competitors, addressing customer needs, and serving as a crucial tool for attracting customers, guiding marketing, and strengthening the startup's competitive edge.
Vesting	Vesting is the process by which equity options are gradually granted to employees or founders over time, typically structured over four years with a one-year "cliff," incentivizing long-term commitment and reinforcing loyalty by requiring individuals to remain with the company to fully benefit from their equity compensation.
Vision Guided	Vision-guided systems use a control mechanism that adjusts a robot's trajectory based on input from a vision sensor, enabling real-time adaptation for tasks like object manipulation, navigation, and automated inspection to enhance accuracy and efficiency.

Visitation Rights	Also known as Observer Rights, visitation rights allow investors to appoint a non-voting representative to attend Board of Directors meetings, ensuring they stay informed about the company's strategy and decisions while fostering transparency, trust, and alignment with leadership.
Voice Processing	Voice processing in AI involves converting spoken language into text (speech-to-text) and generating spoken output from text (text-to-speech), enabling natural interactions between users and AI systems, as seen in virtual assistants like Alexa and Siri, while enhancing applications such as customer service automation and accessibility tools.
Warrant	A warrant is a financial instrument that gives investors the right to purchase a company's stock at a predetermined price in the future, incentivizing funding by allowing investors to benefit from potential stock price appreciation, manage risk, and maximize returns with longer expiration periods and strategic exit options.
Warrant Coverage	Warrant Coverage refers to the issuance of warrants to lenders as a reward for taking on lending risk, aligning their interests with the company's success by allowing them to purchase shares based on a percentage of the loan amount, incentivizing them with potential returns if the company's stock value increases.
Washout Round	A washout round is a financing round where previous investors, founders, and management face significant equity dilution as new investors acquire majority control, often due to financial challenges, diminishing the influence of existing stakeholders while enabling new investors to implement strategic changes.
Waterfall	The waterfall methodology is a structured project management approach used in startups, guiding development through sequential phases, and in liquidity events like mergers or acquisitions, it dictates the distribution of proceeds among stakeholders, prioritizing debts and preferred stockholders before allocating funds to common stockholders, ensuring clarity and fairness in both execution and financial outcomes.
Weak AI	Weak AI, or narrow AI, refers to systems designed to excel at specific tasks within a limited context, such as language translation or image recognition, but lacks the generalized intelligence and adaptability of strong AI, remaining confined to the parameters set by its programming and training.
Weak-to-Strong Generalization	Weak-to-strong generalization is an AI training method where simpler models guide more advanced ones, enabling them to generalize beyond narrow datasets and improving adaptability and performance in broader contexts, such as in reinforcement learning where basic models help stronger models achieve higher performance.

Weighted Average	Weighted average anti-dilution protection adjusts the conversion ratio when new shares are issued at a lower price, with the Narrow-Based Weighted Average offering greater protection to investors by considering only outstanding preferred shares, while the Broad-Based Weighted Average, which includes all fully diluted shares, provides a more balanced outcome slightly favoring the company.
Work Envelope	The work envelope, or workspace, refers to the three-dimensional volume within which a robot or manipulator can operate, defining its range of motion, with its shape and extent influenced by the manipulator's design and positioning, which may impose limitations on reach and flexibility.
Work Piece	A work piece is any material or component undergoing processing, refinement, or manufacturing before it becomes the final product. It represents an intermediate stage in the production cycle, where raw or semi-finished materials are shaped, altered, or assembled into their finished form.
World Model	A world model is a dynamic, three-dimensional representation of a robot's environment, continuously updated by its sensors and stored in memory, enabling the robot to navigate and interact intelligently with its surroundings to complete tasks with precision and adaptability.
Wrist	The wrist is a series of rotary joints connecting the robot's arm to its end-effector, providing multiple degrees of freedom—such as roll, pitch, and yaw—to enable precise orientation and fine-tuned control for manipulating and grasping objects in various orientations.
Write-Off	A write-off is the formal accounting action of reducing the recorded value of an asset or company, typically due to impairment, obsolescence, or uncollectibility.
Yaw, Pitch, and Roll	Yaw, pitch, and roll are key rotational movements in robots and aircraft: yaw is side-to-side rotation, pitch is up-and-down, and roll is tilting, all ensuring precise control of orientation and movement.
Zero-Shot Learning	Zero-shot learning is a machine learning technique that enables models to identify and classify new, unseen concepts by generalizing from related data without needing explicit training on labeled examples.

INDEX

Pages in *italics* refer to figures, pages in bold refer to **tables**, and pages followed by "n" refer to notes.